Standard Operating Procedures for Better Pig Research (SOPig)

Liza R. Moscovice · Sandra Düpjan
Marion Girard · Rebecka Westin

Editors

Standard Operating Procedures for Better Pig Research (SOPig)

Recommendations from the PIGWEB Collaboration

Editors
Liza R. Moscovice
Research Institute for Farm Animal Biology
Dummerstorf, Germany

Sandra Düpjan
Research Institute for Farm Animal Biology
Dummerstorf, Germany

Marion Girard
Swine Research Group, Agroscope
Posieux, Switzerland

Rebecka Westin
Department of Applied Animal Science
and Welfare
University of Agricultural Sciences
Skara, Sweden

ISBN 978-3-032-24898-5   ISBN 978-3-032-24899-2   (eBook)
https://doi.org/10.1007/978-3-032-24899-2

EU Horizon 2020 This work was supported by EU Horizon 2020.

Photo credits: Geena Cartick, PIGWEB

This Springer imprint is published by the registered company Springer Nature Switzerland AG
The registered company address is: Gewerbestrasse 11, 6330 Cham, Switzerland

If disposing of this product, please recycle the paper.

Acknowledgments

The PIGWEB project has received funding from European Union's Horizon 2020 research and innovation program under Grant Agreement No 101004770.

Contents

Introduction

Sandra Düpjan ⓘ, Marion Girard ⓘ, and Rebecka Westin ⓘ

Abstract

A standard operating procedure (SOP) is a step-by-step instruction manual describing how to perform a specific work routine. The primary objective of SOPs in a research setting is to enhance research quality (replicability, reproducibility and repeatability) by establishing standardised ways of performing specific experimental procedures. SOPs serve as a valuable structure for internal communication and sharing of best practices (i.e. training documents to guide new users), and they facilitate the preparation of an experiment and the subsequent publication process, since they provide all information about the equipment needed and how to carry out the procedure in detail. The use of SOPs also makes it easier to collaborate between research institutes, and the generation of more robust and reliable data facilitates its use and re-use for comparison between studies, such as in meta-analyses. This book includes nine SOPs for specific experimental procedures developed within the PIGWEB project. The topics cover both common procedures performed at most experimental facilities, such as blood sampling, and more novel and minimally invasive research methods, such as saliva sampling. The broader goals of this open-access volume are to disseminate current gold standards for pig research while emphasising the importance of future innovations to promote better animal welfare.

Supplementary Information: The online version contains supplementary material available at https://doi.org/10.1007/978-3-032-24899-2_1.

S. Düpjan
Animal Behaviour and Welfare Working Group, Research Institute for Farm Animal Biology (FBN), Dummerstorf, Germany

M. Girard
Swine Research Group, Agroscope, Posieux, Switzerland

R. Westin (✉)
Department of Applied Animal Science and Welfare, Swedish University of Agricultural Sciences (SLU), Skara, Sweden
e-mail: rebecka.westin@slu.se

Keywords

Standard operating procedure · SOP · Research quality · Pigs

Recent years have seen growing public concern regarding the quality of scientific research in general and especially the reproducibility of results (e.g., Calnan et al. 2024). This also applies to research in the animal sciences. Researchers and working groups around the globe have come up with a range of behavioural, physiological and cognitive indicators of animal welfare, health and nutrition over the years, and a variety of methods for how to assess them in farm animals. However, more often than not, research results are inconsistent across studies (Loss et al. 2021). A potential indicator that was useful for detecting differences between two treatments in one study might not be affected by the same (or similar) treatments in another study. This leads one to question whether the indicator itself is valid or whether the hypothesis regarding the treatment effects on animal welfare, health and nutrition is false.

One possible explanation for inconsistent results in animal research is a lack of standardisation or documentation across labs and other research facilities regarding experimental procedures, analytical approaches or investigated populations. The variability in experimental details can roughly be separated into three main areas: (1) animals, (2) housing and management and (3) methods and execution of experimental procedures (as in how/when sampling is carried out, equipment and type of materials used, calibration of technical devices, handling of samples, etc.). Regarding animals, different breeds differ in production performance and may have varying susceptibility or resilience to stress (Oster et al. 2025; Laghouaouta et al. 2024). Also, within breeds, there are often differences between sexes (van den Broeke et al. 2020) or during ontogenetic development (Prunier et al. 2020). Regarding housing and management, there are significant differences in legislation and regulation between countries and/or regions, for example, regarding the use of farrowing crates vs free farrowing (EFSA AHAW Panel 2022). Differences between production schemes (e.g. conventional vs organic vs welfare labels) or feeding practices (e.g. differences in feeding frequency, group or individual feeding and provision of roughage/organic enrichment) can also introduce variation in results. Considering pig research specifically, there is currently a large variation in animals, management and housing conditions across different pig research facilities, which cannot be completely standardised. To a certain degree, this can be seen as a strength since it represents the even larger variation across farms in different countries, where, ideally, results from research would eventually be put into practice. However, regarding how actual sampling is performed in practice and how samples are measured, more can be done to standardise the methods used and how experimental procedures are executed within and across research facilities, to improve research quality and reproducibility.

A standard operating procedure (SOP) is a step-by-step instruction manual describing how to perform a specific work routine. The primary objective of SOPs in a research setting is to enhance research quality (replicability, reproducibility and repeatability). By establishing a systematic way of performing a task, SOPs ensure that the task is done consistently by all involved persons (Manghani 2011), be it within or across research facilities. Development and implementation of SOPs for specific experimental practices is therefore one way of creating more standardised ways of performing research in order to improve the consistency of research methodologies and enhance the reliability and credibility of research findings. The use of SOPs also makes it easier to collaborate between research institutes, and the generation of more robust and reliable data facilitates its use and reuse for comparison between studies, such as in meta-analyses.

Apart from improving the quality of research, the use of SOPs also improves transparency within an organisation, serves as a valuable structure for internal communication and facilitates the sharing of best practices (Amare 2012). An SOP can also contribute to efficiency by serving as a training document to guide new users (e.g. new colleagues or students) through the process outlined in the SOP (Akyar 2012). For researchers, having an SOP can also facilitate both the preparation of an experiment and the subsequent publication process, since it provides all information needed to report how specific procedures were carried out. Introducing SOPs in animal research is also in line with the current implementations of the 3R principles (replace-reduce-refine), as it should help to reduce the number of experiments required by reducing avoidable variability in experimental designs. Furthermore, it should also help with the refinement of methodologies, especially those used across many research groups, by defining best practices to minimise animal stress.

In laboratory research environments and commercial labs, the use of standards from the International Organization for Standardization (ISO) for specific analyses is already routine. However, within many research farms and other facilities designated for research on farm animals, SOPs are currently lacking. In a large collaborative project called PIGWEB (funded by the European Union's Horizon 2020 research and innovation program under Grant Agreement No 101004770), which aimed to connect pig researchers across Europe, we set out to address this lack of SOPs within the leading European pig research infrastructures. The PIGWEB consortium comprises 16 partners from 10 European countries, including 11 research institutions working on pig welfare, health and nutrition. We set a goal of encouraging specialists across a range of institutes and areas of pig research to share and harmonise their existing SOPs as well as to develop new SOPs where needed, for previously undocumented or emerging methods that are important to their own research. This edited volume is the result of those efforts. In general, small teams of pig researchers using the same method in different facilities prepared the initial drafts. These drafts were then internally reviewed by other PIGWEB consortium members who were not specialists in the techniques, allowing them to assess how easy the SOPs were to follow. In addition, every SOP was externally reviewed by researchers outside of the PIGWEB consortium. The final SOPs presented in this book are the outcome of this two-stage review process. Each SOP represents the agreed-upon best practices for carrying out a given procedure with pigs. In addition to emphasising easy-to-follow instructions, the SOPs introduce innovative and minimally invasive techniques whenever possible and emphasise in each protocol best practices for pig welfare, as well as for research purposes. Some of the SOPs have already been used in a multi-lab study (Reimert et al. 2026) in order to assess their applicability. By promoting the dissemination of SOPs within and across pig research facilities, our broader goal is to boost the quality of research by achieving a higher consistency of research methodologies across institutions and having guidelines available to train the next generation of researchers.

This book includes nine SOPs for specific experimental procedures. The topics cover both common procedures performed at most experimental facilities, such as blood, faecal and urine sampling, and more novel and minimally invasive research methods, such as saliva sampling and thermography for physiological assessments, near-infrared scanning/spectroscopy (NIR) to determine ileal digestibility and image analysis to determine body and carcass composition. In addition to the specific experimental procedures presented in this book, PIGWEB has also been promoting the development of SOPs for basic management routines and recording of standard traits in experimental pig research facilities. Such SOPs must, however, be adapted to the specific housing conditions, existing equipment and legislations governing each facility. The results of this work can be found on the PIGWEB website (www.pigweb.eu) and at the Social Science Research Network (SSRN, Westin and Wallenbeck 2024). An example template to be used for the development of internal SOPs for basic management routines is attached in ESM 1.

Several of the SOPs presented in this book emphasise the importance of habituating the pigs to the described sampling techniques and give examples of how this can be done. However, we also want to mention the potential for the use of positive reinforcement training (PRT), also known as clicker training, in pig research. PRT is a growing field within animal research and has successfully been applied in pigs for a range of procedures including the following: (1) intravenous blood sampling without restraint (Fiderer et al. 2024), (2) sling-training (Yang et al. 2021), (3) ultrasound examination (Rydén et al. 2019), (4) urine sampling (Rydén et al. 2019) and (5) blood sampling by catheters without restraint (Rydén et al. 2019). Implementing PRT also provides the animals with cognitive enrichment, which further improves their welfare (Sørensen 2010). Although not the focus of these SOPs, we urge interested readers to inform themselves about PRT and implement such training whenever possible in their own pig husbandry and research procedures.

The huge societal demand for improvements to farm animal welfare and health, and for valid, reliable biomarkers to measure these aspects, should be a motivating factor for all farm animal researchers. This demand can only be met if we improve the quality of our research, raise the standards for best practices, and continue to develop innovative practices reflecting how animals should be farmed in the future.

References

Akyar I (2012) Standard operating procedures (what are they good for?). In: Akyar I (ed) Latest research into quality control. IntechOpen, pp 367–391. https://doi.org/10.5772/50439

Amare G (2012) Review: reviewing the values of a standard operating procedure. Ethiop J Health Sci 22(3):205–208

Calnan M, Kirchin S, Roberts DL et al (2024) Understanding and tackling the reproducibility crisis—why we need to study scientists' trust in data. Pharmacol Res 199:107043. https://doi.org/10.1016/j.phrs.2023.107043

EFSA AHAW Panel (2022) Scientific opinion on the welfare of pigs on farm. EFSA J 20(8):7421. https://doi.org/10.2903/j.efsa.2022.7421

Fiderer D, Thoene-Reineke C, Wiegard M (2024) Clicker training in Minipigs to reduce stress during blood collection—an example of applied refinement. Animals 14(19):2819. https://doi.org/10.3390/ani14192819

Laghouaouta H, Fraile LJ, Pena RN (2024) Selection for resilience in livestock production systems. Int J Mol Sci 25:13109. https://doi.org/10.3390/ijms252313109

Loss CM, Melleu FF, Domingues K et al (2021) Combining animal welfare with experimental rigor to improve reproducibility in behavioral neuroscience. Front Behav Neurosci 15:763428. https://doi.org/10.3389/fnbeh.2021.763428

Manghani K (2011) Quality assurance: importance of systems and standard operating procedures. Perspect Clin Res 2(1):34–37. https://doi.org/10.4103/2229-3485.76288

Oster M, Gladbach CA, Vernunft A et al (2025) Influence of genotype-environment interaction on stress parameters during spontaneous farrowing in modern and traditional pig breeds housed in crates and pens. Theriogenology 240:117394. https://doi.org/10.1016/j.theriogenology.2025.117394

Prunier A, Averos X, Dimitrov I et al (2020) Review: early life predisposing factors for biting in pigs. Animal 14(3):570–587. https://doi.org/10.1017/S1751731119001940

Reimert I, Amorim Franchi G, Bagaria M et al (2026) Teaming up for welfare: a cross-institute approach to identify robust indicators of emotion in pigs. Animal 101871. https://doi.org/10.1016/j.animal.2026.101871

Rydén A, Manell E, Biglarnia A et al (2019) Nursing and training of pigs used in renal transplantation studies. Lab Anim 54(5):469–478. https://doi.org/10.1177/0023677219879169

Sørensen DB (2010) Never wrestle with a pig. Lab Anim 44:159–161. https://doi.org/10.1258/la.2010.00910

van den Broeke A, Leen F, Aluwé M et al (2020) The effect of sex and slaughter weight on performance, carcass quality and gross margin, assessed on three commercial pig farms. Animal 14(7):1546–1554. https://doi.org/10.1017/S1751731119003033

Westin R, Wallenbeck A (2024) Protocols for standard management and recording in pig research facilities. Available at SSRN: https://ssrn.com/abstract=4750460

Yang HY, Galang KG, Gallegos A et al (2021) Sling training with positive reinforcement to facilitate porcine wound studies. JID Innov 1(2):100016. https://doi.org/10.1016/j.xjidi.2021.100016

The Usefulness of Standardised Ontologies for Livestock Research

2

Catherine Hurtaud, Etienne Labussière, Alice Fatet,
Marie-Madeleine Mialon, Romain Lardy,
Matthieu Reichstadt, and Jérôme Bugeon

Abstract

We introduce the Animal Trait Ontology for Livestock (ATOL) and the Environment Ontology for Livestock (EOL), which were built by INRAE researchers to provide a systematic and interoperable tool for describing traits across species and animal production systems. The aims of ATOL and EOL are to promote the establishment of a common language between zootechnicians, physiologists and geneticists to facilitate data sharing across projects, institutions and countries, particularly within large-scale consortia and international databases. Standardised ontologies also reduce the need to repeat experiments, save time and resources and encourage open science by making data FAIR (Findable, Accessible, Interoperable and Reusable). The SOPs in this volume have been annotated using ATOL and EOL to facilitate quick retrieval of different SOPs related to various phenotypic traits. The ATOL traits fall within the broad categories of growth and meat production, fatty liver, mammary gland and milk production, egg, nutrition, reproduction and animal welfare. The EOL traits fall within the broader categories of livestock feeding, livestock farming structure, livestock farming environment and livestock farming systems. A full index of the ATOL and EOL terms can be found in the index.

C. Hurtaud (✉) · E. Labussière
PEGASE, INRAE, Institut Agro, Saint Gilles, France
e-mail: catherine.hurtaud@inrae.fr

A. Fatet
INRAE, FERLUS, Lusignan, France

M.-M. Mialon · R. Lardy
Université Clermont Auvergne, INRAE, VetAgro Sup, UMR Herbivores,
Saint-Genes-Champanelle, France

M. Reichstadt
INRAE, CIRAD, L'Institut Agro, SELMET, Montpellier, France

J. Bugeon
INRAE, LPGP, Rennes, France

© The Author(s) 2026

L. R. Moscovice et al. (eds.), *Standard Operating Procedures for Better Pig Research (SOPig)*, https://doi.org/10.1007/978-3-032-24899-2_2

Keywords

Phenotype · Trait · Standardisation · FAIR data · Production systems

Animal phenotyping—the characterisation of an animal's observable traits—is critical for livestock research, especially related to developing breeding solutions to improve animal performance and for animal physiology research. However, phenotypic data are often heterogeneous, fragmented and insufficiently annotated. To address this, the development and application of structured ontologies are necessary. An ontology is a formal representation of a set of concepts and the relationships between these concepts in a specific domain (Bard and Rhee 2004). Animal Trait Ontology for Livestock (ATOL) and Environment Ontology for Livestock (EOL) are ontologies built by INRAE researchers that offer systematic and interoperable tools for describing traits across species and animal production systems (https://www.umrh.inrae.fr/ontologies/visualisation/public/). The aim of ATOL and EOL is to promote the establishment of a common language between zootechnicians, physiologists and geneticists to facilitate collaborative projects between different disciplines and/or animal models and to make information shareable through the use of universally recognised characters that are referenced in publications and databases. A particular effort was made to organise the ontology as a specialisation hierarchy, where each class (or N-level sub-branch) is a specialisation (or subclass, child class or N-1 level sub-branch) of its superclasses. ATOL has seven main classes: growth and meat production, fatty liver, mammary gland and milk production, egg, nutrition, reproduction traits and animal welfare traits. EOL has four main classes: livestock feeding, livestock farming structure, livestock farming environment and livestock farming system.

One of the benefits of ATOL and EOL is through greater standardisation. Phenotypic traits are typically recorded using various terminologies across laboratories, countries and research domains. Ontologies provide a controlled vocabulary that ensures consistency in how traits are described, promoting semantic interoperability among databases and tools. This standardisation is essential for the following: (1) data integration from diverse sources and experimental setups; (2) cross-species comparisons, enhancing translational research between model organisms and livestock; and (3) automated data mining, allowing systems to interpret and process data accurately.

With ATOL and EOL, traits are clearly defined with unambiguous relationships and descriptors. Such systematic ontologies facilitate data reuse by making phenotypic information machine-readable and interoperable. This enhances data sharing across projects, institutions and countries, particularly within large-scale consortia and international databases. Such reuse has practical implications. For example, researchers can combine datasets for meta-analyses. Standardised ontologies also reduce the need to repeat experiments, saving time and resources and ontologies encourage open science by making data FAIR (Findable, Accessible, Interoperable and Reusable).

ATOL and EOL enable rich annotation of phenotyping data. Each trait can be linked with biological processes, measurement methods, environmental conditions and anatomical references. This enhances semantic search capabilities. For instance, a query for "growth traits in pigs" can retrieve relevant traits such as "live weight", "average daily gain" and "body length," even if they were recorded under different terminologies. By enabling structured representation of traits, ontologies improve database quality and accuracy and thus facilitate modelling in livestock systems (Hurtaud et al. 2023). Specifically, ATOL and EOL support automated reasoning and inference, which can help to identify hidden patterns or biological relationships within complex data sets.

ATOL and EOL are designed not just for one specific research area but can be used in agronomic, veterinary and genomic domains. This cross-domain applicability fosters collaboration and unifies language between professionals from different disciplines. For example, a veterinary clinician and a livestock geneticist can interpret "udder depth" or "lameness" traits using the same definitions. Industry partners can interoperate with research databases more effectively. The ontology also supports multilingual term representation, enhancing accessibility in international contexts. Researchers in non-English-speaking regions can work with trait data in their native languages while maintaining global interoperability.

Each of the SOPs included in this book has been annotated using ATOL and EOL, resulting in the inclusion of unique identifiers for the first occurrences in every chapter of any traits that are listed in the ATOL or EOL databases. The complete list of these traits, with their associated ATOL or EOL descriptors, is listed in the index.

2.1 Conclusions

Ontologies like ATOL and EOL can improve animal phenotyping by introducing clarity, interoperability and automation to a domain historically plagued by inconsistency and fragmentation. It empowers data-driven decision-making, supports cutting-edge research and fosters international collaboration. As the livestock sector increasingly turns to big data and precision agriculture, the role of ATOL and EOL in harmonising phenotypic knowledge will become indispensable.

References

Bard JBL, Rhee SY (2004) Ontologies in biology: design, applications and future challenges. Nat Rev Genet 5:213–222

Hurtaud C, Bernard L, Boutinaud M et al (2023) Feed restriction as a tool for further studies describing the mechanisms underlying lipolysis in milk in dairy cows. Animal - open space 2:100035

Variation in Blood Sampling Techniques: From Venipuncture to Catheters

Nathalie Le Floc'h, Francis Aman Eugenio, Axel Sannö,
Bjorge Laurensen, Örs Petneházy, Miriama Sciascia,
Andreas Vernunft, Hélène Quesnel,
and Catherine Ollagnier

Abstract

Many experimental protocols require analytes to be measured in blood. However, collecting blood from pigs is difficult because they have few superficial vessels that are easily accessible. Therefore, it is important to consider whether the parameter can be measured in another easily collected matrix (e.g. saliva). If blood is required, the choice of blood collection method and site should be based on the volume of blood needed and the sampling frequency. In

Supplementary Information: The online version contains supplementary material available at https://doi.org/10.1007/978-3-032-24899-2_3.

N. Le Floc'h (✉) · H. Quesnel
PEGASE, INRAE, Institut Agro, Saint Gilles, France
e-mail: nathalie.lefloch@inrae.fr

F. A. Eugenio · C. Ollagnier
Swine Research Group, Agroscope, Posieux, Switzerland

A. Sannö
Department of Clinical Sciences, Swedish University of Agricultural Sciences (SLU),
Uppsala, Sweden

B. Laurensen
Department of Animal Science, Adaptation Physiology Group, Wageningen University and
Research (WUR), Wageningen, The Netherlands

Ö. Petneházy
Medicopus, 7400, Aposvár, Hungary

M. Sciascia
Nutritional Physiology Working Group, Research Institute for Farm Animal Biology (FBN),
Dummerstorf, Germany

A. Vernunft
Animal Husbandry & Health Working Group, Research Institute for Farm Animal Biology
(FBN), Dummerstorf, Germany

L. R. Moscovice et al. (eds.), *Standard Operating Procedures for Better Pig
Research (SOPig)*, https://doi.org/10.1007/978-3-032-24899-2_3

practice, however, not all procedures have the same consequences for the animal. Some involve short-term pain (e.g. when the animal is restrained with a snout rope), while others involve medium-term stress (e.g. when individual housing is imposed). These aspects must also be considered when choosing the most appropriate blood collection method. This standard operating procedure describes the different blood collection techniques available in pigs and provides guidance for readers on choosing the most appropriate technique according to their needs.

Keywords

Blood vessels · Blood collection · Pain · Restraining · Training · Umbilical cord

3.1   Introduction

Sampling blood (ATOL_0005631) is a routine procedure used to assess the health (ATOL_0000928) of pigs (as part of a veterinary examination) or to measure blood parameters in scientific experiments. Blood can indeed be used to measure a variety of molecules, including hormones (ATOL_0000527), nutrients, metabolites and antibodies, and to count and collect blood cells for culture or measurement of gene expression. When blood sampling is performed for experimental purposes, it must comply with the regulations on the protection of animals used for scientific purposes (Directive 2010/63/EU). Unlike horses or cattle, pigs have very few superficial blood vessels that are easily accessible. Furthermore, pigs can be fearful and are strong animals, the approach of untrained personnel can therefore be dangerous when collecting blood. Appropriate restraint of the pigs during collection is therefore required. This makes blood collection quite stressful for the pigs, especially when repeated collections are needed. Consequently, the procedure requires a high level of skill, extensive training and regular practice for those performing the collection.

As laboratory analysis techniques have evolved, it has become possible to measure certain molecules and chemical properties in a small sample of blood, such as a drop. Minimally invasive sensors are also an interesting alternative. While these methods show promise, they still require testing for development and implementation in farmed animals. For this reason, this SOP will focus only on the most commonly used blood collection methods for pigs. Several techniques for blood collection are available, depending on factors such as the size of the pig and the collection site (blood vessel). As blood collection is labour-intensive and requires specific training, having a reference document describing the different steps of the activity benefits both the animal and the personnel.

3.2   Goals and Scope

The aim of this SOP is to describe the best methods for collecting blood samples from pigs of all ages. The objective is to provide guidance on choosing techniques and sampling sites adapted to the age and body weight (ATOL_0000351) of the pigs, while constantly aiming to minimise stress (ATOL_0002301) and pain (ATOL_0000863) and guarantee the quality of the collected blood. The surgical procedures for catheter insertion and the management of samples after blood collection and before laboratory analysis are not covered.

3.3   Operator Training and Prerequisite

All personnel involved in any blood procedures should have received prior training on animal handling and restraining. The personnel collecting blood should be well-trained and also have sufficient knowledge of the pig's anatomy. Hence, inexperienced personnel should always be supervised. Blood sampling requires the pig to be restrained to ensure that it remains still during the procedure. This is as important for the pig as it is for the safety of personnel taking the blood sample, especially when blood needs to be collected in non-anaesthetised pigs heavier than 30 kg. Pigs have to be handled with care and respect.

The personnel should preferably wear a noise-cancelling device and medical gloves or make sure that their hands are clean. For venipuncture, the site of puncture should be cleaned first and ideally disinfected. Always use sterile and single-use materials (i.e. syringe, needles, indwelling catheters, etc.) that should not be reused between animals. This ensures proper hygiene during collection, which is mandatory and should be observed at all times to prevent infections. In addition, special attention should be paid to have enough light to see the vessels for blood sampling.

The list of materials and equipment required for all procedures should include the appropriate vacuum tubes or syringes for collecting blood. Vacuum tubes and syringes must be used with a sterile needle, the size and diameter of which should be adapted to the size of the pig (see Supplementary Table 3.1). Vacuum tubes have two main advantages. Firstly, they can be prepared with different additives to prevent blood clotting. Blood can remain in a vacuum tube for a few hours before being transferred to the laboratory. The additive must be chosen according to the planned analyses. The second advantage is that the tube automatically fills when the needle reaches the blood flow, eliminating the need for further manipulation. This minimises the risk of the needle coming out or the vessel being damaged. When using a syringe, it is possible to manually control the vacuum and the volume and rate of blood collection. Once collected, the blood must be transferred to a tube containing the appropriate additive. Tubes filled with blood must be kept in the appropriate conditions until transfer (e.g. an ice box or tube rack), and the following items must be available: a needle holder for procedures requiring venipuncture with a vacuum tube, examination gloves, absorbent paper or a clean cloth to clean blood from materials and a medical bin for any materials stained with blood. For materials specific to each procedure, refer to the Supplementary Table 3.1.

3.4   Recommendations for Choosing the Blood Sampling Technique

The main criteria influencing the choice of blood sampling technique include the size of the pig, the sampling site, frequency and the required blood volume (see Supplementary Table 3.1). In order to minimise the negative effects of blood sampling on the health and welfare (ATOL_0000765) of pigs as much as possible, the maximum volume that can be collected within a given period should be defined.

Important

In line with the established guidelines adopted by most countries (Diehl et al. 2001; Wolfensohn and Lloyd 2003), the maximum blood volume per kg of body weight that can be collected from pigs can be estimated from the following principles:

- Total blood volume (TBV) of a swine is about 60 mL/kg body weight (BW) or 6.0% of total BW.
- No more than 1.0% (0.6 mL/kg BW) of TBV in one collection over a 24-h period.
- No more than 7.5% (4.5 mL/kg BW) of TBV can be collected in a single or multiple draws over a week period.
- If 10% (6 mL/kg BW) of total blood volume is collected, at least 10 days of recovery must be allowed before the next draw.
- If 15% (9 mL/kg BW) of total blood volume is collected, at least 15 days of recovery must be allowed before the next draw.

3.5   Blood Collection from the Umbilical Cord of Newborn Piglets

3.5.1   Vessel Location and Techniques

The umbilical cord contains three vessels: two arteries and one vein (Fig. 3.1). As umbilical cords do not contain nerves, umbilical blood sampling is not considered painful for animals. While restraining them may cause some discomfort, newborn piglets are not very mobile or responsive. Blood can be collected by venipuncture, from an intact or broken cord, or by 'dripping' from a broken cord only. Compared to venipuncture, the dripping technique does not require maintenance of blood pressure in the umbilical cord, making it more reliable and less prone to failure. However, this technique mixes blood from the arterial and venous systems, which may not be appropriate for measuring certain parameters. Meanwhile, venipuncture allows for more accurate sampling of venous blood (veins are larger and have thinner walls), but it is not possible to exclude arterial samples.

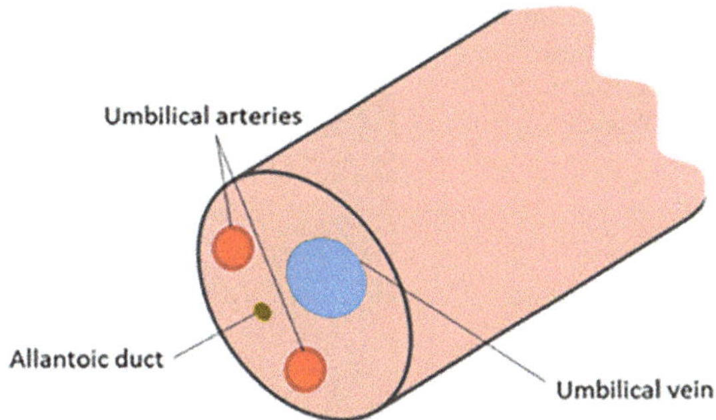

Fig. 3.1 Cross-sectional anatomy of the piglet's umbilical cord (Photo credit: Francis A. Eugenio, Agroscope)

3.5.2 Pig Handling and Care

Operators should be present during farrowing to ensure that blood samples can be taken immediately after the piglet is expelled. Collecting blood by venipuncture from an intact umbilical cord does not require the piglet to be handled. This method limits contact between humans and piglets.

However, in most cases, newborn piglets break their umbilical cords themselves, leaving them between 20 and 40 cm long. If the umbilical cord is intact, measure 5 cm from the piglet and tear it by hand (do not use a knife or scissors; tearing allows the cord to seal more quickly). Blood can then be collected by dripping or venipuncture. Then, wrap the piglet in a tissue or clean cloth and pass it to a colleague. Ask them to hold the piglet from underneath with one hand behind the front legs and the other in front of the back legs, ensuring that the umbilical cord is held firmly between both hands. This step should be relatively easy due to the piglet's size and low mobility soon after birth, but make sure not to strain them excessively or cause any unnecessary discomfort. Clamps and tourniquets should always be available and ready for each piglet in case of bleeding.

3.5.3 Description of the Procedures

3.5.3.1 Venipuncture from Intact or Broken Umbilical Cord
The vein should have been properly located before any venipuncture. Then proceed as follows:

1. Immobilise the cord between two fingers and place a third finger below the cord to prevent it from rolling. The cord will be punctured just above the third finger (Fig. 3.2).
2. Puncture the umbilical vein at an angle of 30°. Then, lower the needle slightly parallel to the cord and penetrate the vein gently.
3. Draw a maximum of 3 mL of blood and then withdraw the needle gently.
4. Clamp the umbilical cord with an umbilical clamp (or a surgical bandage) placed near the piglet's umbilicus to prevent any bleeding.

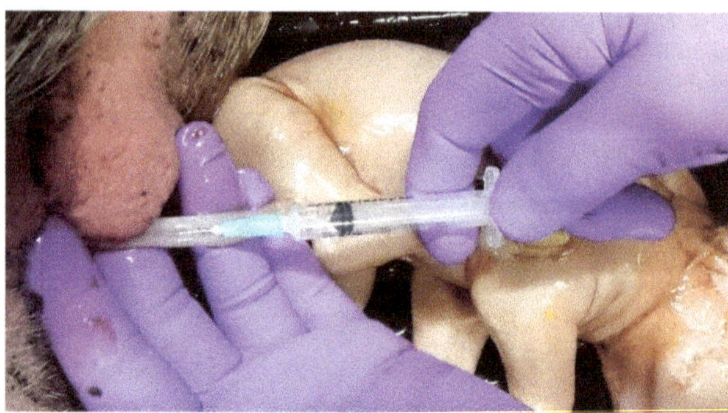

Fig. 3.2 Blood sampling in the umbilical cord (Photo credit: Bjorge Laurensen, WUR)

3.5.3.2 Dripping

1. Use two fingers to seal or clamp the end of the umbilical cord.
2. Open the collection tube and hold it with one hand, aligning it to the umbilical cord.
3. Slowly release the finger clamp. If the blood does not come out or is not flowing enough, take your fingers and place them at the junction between the umbilical cord and the piglet.
4. Gently apply pressure and 'milk' downwards towards the tube. Repeat the milking steps until the sample volume required (3 mL maximum) has been collected.

3.5.4 Post-procedure Observation

If the umbilical cord continues to bleed, use a tourniquet (such as an umbilical clamp or surgical bandage) to tie it off. Remove the clamp or bandage after a few minutes to allow the piglet to move freely.

3.6 Blood Collection by Venipuncture of Veins Accessible by the Fossa Jugularis

3.6.1 Puncture Location

The most common technique for collecting blood from pigs involves puncturing one of the large veins located within the jugular groove on the ventral side of the neck. The largest of these veins is the anterior vena cava, which is very close to the heart and receives blood from the jugular veins and the two brachiocephalic veins (see Fig. 3.3). This technique has been widely practised and established and can be applied to pigs of varying sizes. However, it is not possible to visualise the veins, so venipuncture relies exclusively on anatomical landmarks. Consequently, it is difficult to know exactly which vein will be punctured. Moreover, the technique is easier to apply to awake pigs. While it is possible to collect blood from anaesthetised pigs,

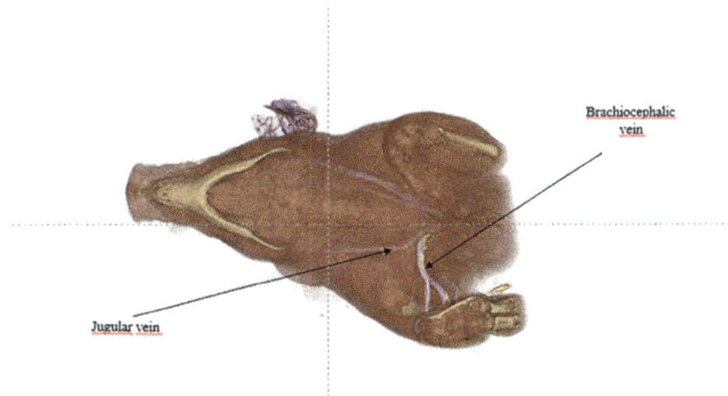

Fig. 3.3 A volume rendered image from a CT scan of a piglet showing the position of the jugular and brachiocephalic veins (Photo credit: Örs Petneházy, Medicopus)

the vessel will be difficult to access due to muscle relaxation. In that case, it may be advisable to use an echograph to locate the vein.

3.6.2 Pig Handling

The technique and process for collecting blood from the jugular vein or anterior vena cava is similar regardless of the size of the pig. The main differences lie in the position of the pig and the method of restraint.

For pigs weighing less than 20 kg, particularly piglets, it is possible to restrain them by hand. However, restraining tools such as a V-trough, a halved tube or a pig sling may also be used for more convenient sampling (Fig. 3.4). When using a V-trough or a tube, we recommend placing a cushion under the pig's back. If using a pig sling, ensure that the pig is properly fitted into the device (i.e. that its feet are well placed within the holes). Ideally, the ventral side of the neck should be easily accessible for puncture. Otherwise, tilt the head back to gain proper access to the puncture site. When using another restraint tool, personnel should gently hold the pig in a supine position with the forelegs pointing towards the rear of the animal, extending the neck with one hand while holding it in place with the other. It may also be helpful to have another person hold the pig's hind legs. This position is uncomfortable for the pig, so restraining for no more than 2 minutes is recommended. Unsuccessful sampling should be expected. Once restrained, the sampling site should be cleaned with water at least, and disinfection with alcohol or another disinfecting solution is strongly advised.

For pigs weighing over 20 kg, it is often necessary to restrain them using a snout rope around the upper jaw (Fig. 3.5). The snout rope should be placed tightly around the upper jaw, behind the canines and held taut. When restrained with the snout rope, the pig should stand on all four legs with an upright body and slightly stretched neck. To avoid straining the pigs excessively, ensure their forefeet remain on the ground. Only personnel who have received sufficient training in the use of the snout rope should do so, as this can be very painful for the pig. For this reason, we recommend not restraining the pig for more than 2 minutes. Unsuccessful sampling should therefore be expected.

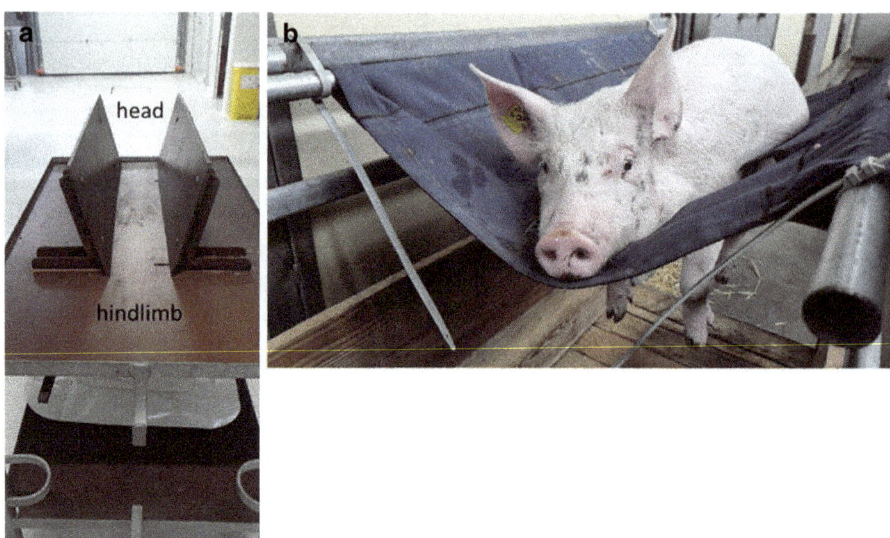

Fig. 3.4 Examples of restraining tools that can be used for piglets and small pigs (< 20 kg) during venipuncture of the jugular vein/anterior vena cava. (**a**): Adjustable V-shaped restrainer. (**b**): Pig sling (Photo credit: Josselin Delamarre, INRAE)

Fig. 3.5 A snout rope that is used to restrain pigs heavier than 20 kg (Photo credit: Catherine Ollagnier, Agroscope)

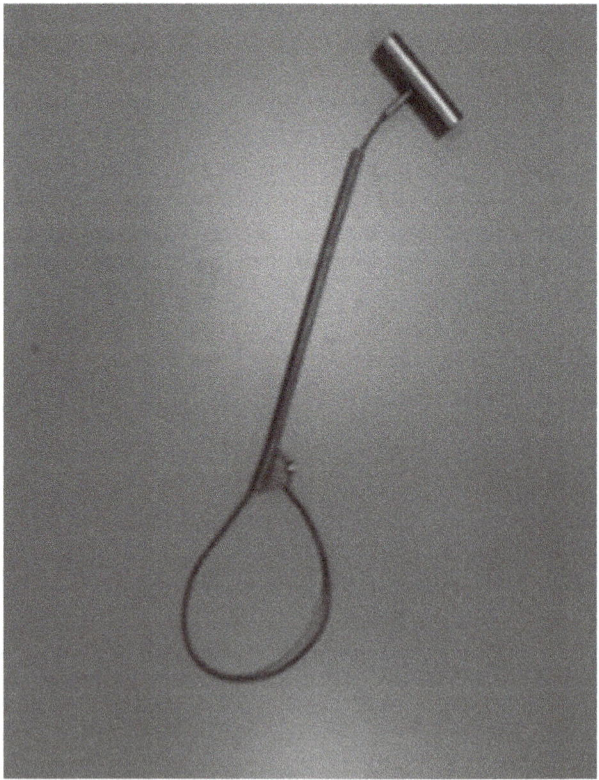

It is possible to train pigs to collect blood without restraining them using positive reinforcement training techniques, as demonstrated by Fiderer et al. (2024). Positive reinforcement training (PRT), also known as clicker training, involves short, multi-step sessions with gradually increasing challenges. Highly palatable rewards such as sugar water or fruit are given when the pig displays the desired behaviour of standing still and stretching its neck voluntarily. PRT has also been used in pigs for sling training (Yang et al. 2021) and for ultrasound examinations, urine sampling and blood sampling via catheters without restraint (Rydén et al. 2019).

PRT has been shown to effectively reduce stress during blood sampling in pigs, resulting in lower salivary cortisol levels (ATOL_0005349) and a slower heart rate (ATOL_0000796) after sampling compared to before training (Fiderer et al. 2024). This improves the quality of research. PRT also improves working conditions for staff and provides cognitive enrichment for animals, thereby promoting welfare. It is therefore strongly recommended in all experimental settings (Sørensen 2010).

3.6.3 Description of the Procedure

3.6.3.1 Locating the Puncture Site

- For pigs under 20 kg: the preferred puncture site is in the middle of the fossa jugularis, cranial to the tip of the sternum. Usually 3–4 cm cranially to the sternum and lateral to the trachea (Fig. 3.6).
- For pigs heavier than 20 kg: the preferred puncture site is in the middle of the fossa jugularis, cranial to the tip of the sternum. Depending on the size of the pig, this can be found 3–10 cm cranially and laterally from the sternum, in the deepest part of the fossa.

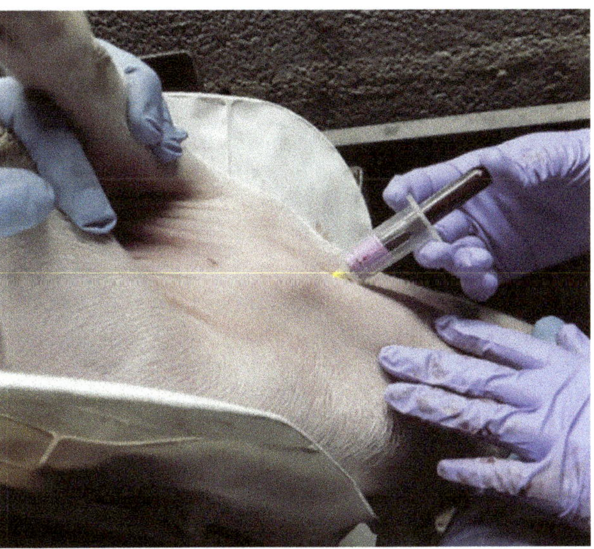

Fig. 3.6 Venipuncture of the jugular vein/anterior vena cava for piglets and small pigs (< 20 kg). The needle is placed at the jugular fossa between the sternohyoideus and brachiocephalicus muscles. The needle points to the direction of the opposite coxal tuberosity (Photo credit: Nathalie Le Floc'h, INRAE)

- Preferably, do not use the left side to avoid vagal nerve damage.
- Clean the general area of the puncture site (at least with water). Disinfection with any disinfecting solution is advised.

3.6.3.2 Puncture of the Jugular Vein

1. Penetrate the skin (ATOL_0005636) with the tip of the needle while holding the vacuum tube in case of using the vacutainer technique. Connect the vacuum tube to the needle only once the needle penetrates the skin to keep the vacuum in the tube. If using a syringe, pull the plunger of the syringe as soon as the needle penetrates the skin to create a vacuum.
2. Push the needle deeper, keeping it perpendicular to the skin or pointing to the opposite coxal tuberosity.
3. Immediately stop pushing the needle when blood appears and freely flows into the tube or syringe.
4. When using a vacuum tube for sampling, multiple tubes can be filled in one venipuncture.
5. If you have missed the vein: Reposition the needle carefully, without exiting fully back out through the skin, and try puncturing again (maximum of 3 attempts as recommended by Parasuraman et al. 2010 and Bollen et al. 2010).
6. Once the required volume of blood is collected, stop pulling on the plunger of the syringe or immediately detach the vacuum tube.
7. Carefully remove the needle and immediately apply pressure in a circular motion for around 30 sec on the puncture site after it exits the skin. Clean blood on skin, if any.
8. If using a syringe, immediately transfer the blood sample from the syringe into the appropriate sampling tube before it coagulates in the syringe.
9. Manually turn the tube four to five times to let the blood mix with the coat additives on the walls of the tube.
10. Label the tube.

3.6.4   Post-procedure Observation

Check the puncture site for any bleeding or bruising. If there is excessive bleeding, apply pressure to the puncture site for a couple of minutes. Applying ice can help to stop the bleeding. When returned to the pen, the pig should be moving correctly with a proper gait.

Young pigs weighing less than 10 kg may experience vagal malaise (i.e. syncope, convulsions, hypersalivation and cyanosis) during or after blood collection (Straw et al. 2006). In the vast majority of cases, this malaise is reversible. The pig should be isolated in a quiet, heated place and monitored until it returns to normal behaviour (30 min to 1 h). Once the pig is active again, it can be returned to its regular pen or cage.

If a snout rope was used, check for possible snout injury or broken tooth and provide appropriate care if needed.

Remarks:

- Puncture of an artery is possible. In that case, the collected blood is bright red instead of being dark red for venous blood.
- If the pig is very mobile during venipuncture, blood collection may not be feasible. In such cases, withdraw the needle and stop the sampling because of the risk of tearing the vessel, which can be serious when the vessel is an artery.
- If the blood stops flowing during the sampling:
  - Check whether the head is properly fixed straight.
  - The tip of the needle may be against the wall of the vein. We then recommend gently twisting the needle to uncover the opening.

  Veins can collapse due to the vacuum applied during puncture. Release the vacuum (remove the tube from the holder, or release the plunger) to reopen the vein. Blood can be drawn again by reapplying a vacuum.
- Stop the procedure when the time to restrain the pig and to collect blood exceeds 5 mins for pigs manually restrained and 2 min when using a snout rope.
- Unsuccessful sampling should be expected and accepted.

3.7   Blood Collection from Cephalic or Saphenous Veins (Pigs Heavier than 20 Kg)

3.7.1   Puncture Site Location and Pig Handling

The cephalic vein runs superficially on the cranial side of the forelegs, and the saphenous vein runs superficially on the cranial side of the hind legs, respectively, in pigs. These medium-sized vessels can withstand a maximum volume of 5 mL being withdrawn without collapsing. For the cephalic vein, the puncture site is on the caudomedial surface of the carpus (Fig. 3.7a). For the saphenous vein, the puncture site is on the lateral surface of the tibia (Figure 3.7b). Clip the site to improve visibility of the vein.

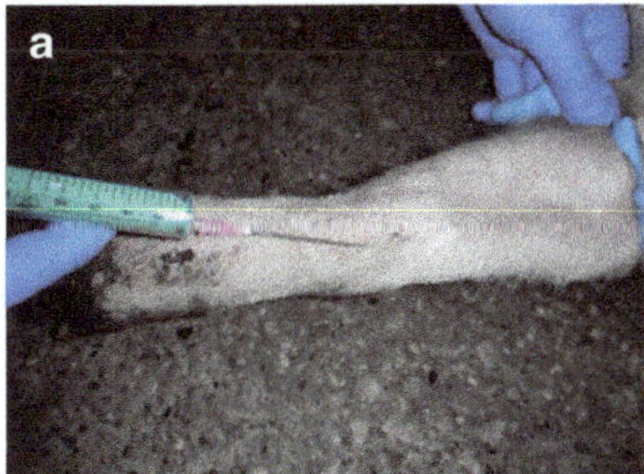
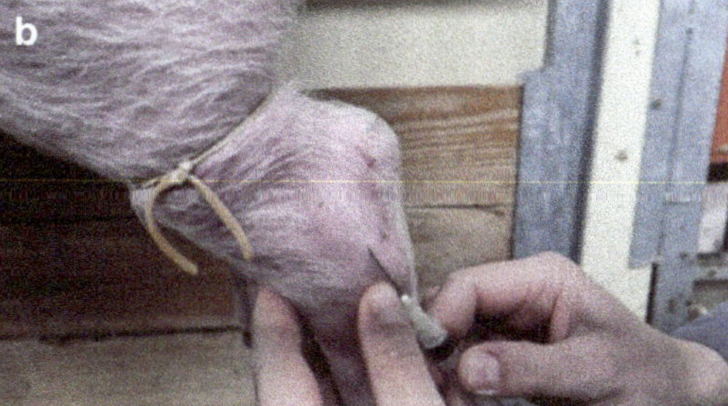

Fig. 3.7 Blood collection from the cephalic vein in a 4-month-old pig (**a**) and from the saphenous vein (**b**) (Photo credit: Catherine Ollagnier, Agroscope)

Anaesthesia is preferable for this method, as restraining the pig's feet while it is awake is difficult. The pig should be placed in lateral recumbency. When collecting blood from the cephalic or saphenous veins, the lower part of the leg (below the knee or elbow) is the preferred location for this technique. This allows for a more consistent anatomical landmark and uses veins that are closer to the skin's surface than in the upper part of the leg. When working with awake pigs, a snout rope can be used to restrain them. If possible, awake pigs should be placed in lateral recumbency. Additional personnel may be needed to hold the pig's feet. Either of these positions may distress the pig, so restraint should not exceed 5 min. Another effective restraint method is the use of a pig sling (Fig. 3.4b). Note that blood can easily be collected from the saphenous vein (hind leg) in unanaesthetised boars during semen collection when the boar is standing on its hind legs (Ferchaud et al. 2019).

3.7.2 Description of the Procedure

1. The backflow of the vein should be restrained at the elbow/knee joint using a tourniquet.
2. Visualise/palpate the vein.
3. Clean the general area of the puncture site, (at least with water). Disinfection with any disinfecting solution is advised.
4. Puncture the vein and drive the needle towards the direction of the vein, flat under the skin. Remove the tourniquet.
5. After penetrating the skin with the needle, apply a small vacuum by slightly pulling on the plunger of the syringe or when working with vacuum tubes, attach the tube to the needle hub. Then begin carefully pushing the needle deeper.
6. Once blood freely flows into the syringe or vacuum tube, immediately stop pushing the needle and collect the needed amount of blood.
7. Once the required volume of blood is collected (maximum 5 mL), stop pulling on the plunger of the syringe or immediately detach the vacuum tube.
8. Carefully remove the needle and immediately apply pressure in a circular motion for around 30 sec on the puncture site after it exits the skin. Clean blood on skin, if any.
9. If using a syringe, immediately transfer the blood sample from the syringe into the appropriate sampling tube before it coagulates in the syringe.
10. Manually turn the tube four to five times to let the blood mix with the coat additives on the walls of the tube.
11. Label the tube

3.7.3 Post-procedure Observation

Check the puncture site for any bleeding or bruising. If there is excessive bleeding, apply pressure to the puncture site for a couple of minutes. When returned to the pen, the pig should be moving correctly with a proper gait. If the procedure was performed on anaesthetised pigs, they should only be returned to the pen once they have fully regained consciousness, appear healthy and are moving well. If a snout rope was used, check for possible snout injury.

3.8 Blood Collection from Mammary (Cranial Epigastric) Veins in Pigs Heavier than 30 Kg and Lactating Sows

3.8.1 Puncture Site and Pig Handling

This blood collection technique is suitable for large pigs, as the mammary veins are large enough to be punctured. Like the cephalic and saphenous veins, the mammary veins are superficial and can be seen with sufficient dilation in non-lactating animals. However, for lactating sows, the veins cannot be clearly seen as they are hidden by the mammary gland. Nevertheless, blood can be collected without the use of a snout rope when the sows are in a farrowing crate, making the procedure less invasive for these animals (Scollo et al. 2019).

For growing pigs, this technique should be performed on anaesthetised animals. The pig should be held lying on its back. Collection is performed on the ground, but the use of a V-trough is preferred (Fig. 3.4a). Depending on the pig's size, one or two people should hold its legs to open up access to its belly/vein for the person doing the collection.

For lactating sows, this technique can be performed without anaesthesia or further restraint when the sow is in a farrowing crate. Sampling can be performed when the sow is standing, sitting, lying down or nursing. In this case, it may be possible for a single person to perform the procedure.

3.8.2 Description of the Procedure

3.8.2.1 For Growing Pigs
1. Locate the end of the costal arch close to the sternum (Fig. 3.8).
2. Gently press the area after the last costal rib cartilage close to the xyphoid cartilage to restrain the backflow on the cranial epigastric vein.
3. Try to dilate the blood vessels by rubbing them with a swab soaked in warm water.
4. Clean the general area of the puncture site (at least with water).
5. Disinfection with any disinfecting solution is advised.
6. Only perform puncture on superficial veins that are clearly visible to avoid repeated and unnecessary punctures.

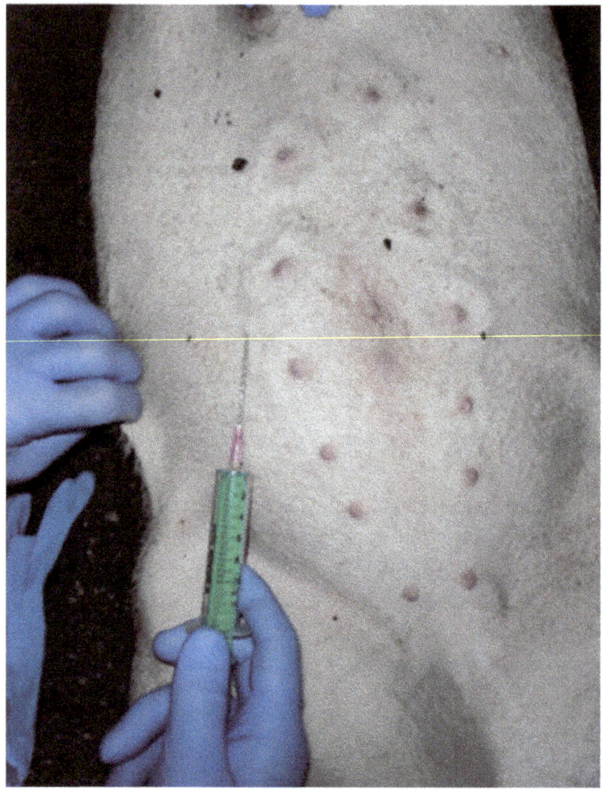

Fig. 3.8 Locating the mammary vein (cranial epigastric vein) in a 4-month-old pig. The assistant's hand is pressing the backflow at the sternum (not visible in the picture) (Photo credit: Catherine Ollagnier, Agroscope)

7. Puncture is made lateral to the mammary glands and towards the head of the pig where you can palpate and see the vein under the skin.
8. When using a syringe, apply a small vacuum shortly after puncturing the skin by gently pulling on the plunger of the syringe. Meanwhile, when using a vacuum tube, connect it immediately via the needle holder after the needle punctures the skin.
9. Push the needle deeper and immediately stop once blood is freely flowing into the tube or syringe.
10. Once the required volume of blood is collected, stop pulling on the plunger of the syringe or immediately detach the vacuum tube.
11. Carefully remove the needle and immediately apply pressure in a circular motion for around 30 sec on the puncture site after it exits the skin. Clean blood on skin, if any.

3.8.2.2 For Lactating Sows

1. Be aware that the vein is not visible, but no vessel pressure is necessary as the blood flow in the mammary vein is sufficient to perform blood collection. Venipuncture is performed between the second and third anterior gland, in the external side of the mammary gland (Fig. 3.9a).

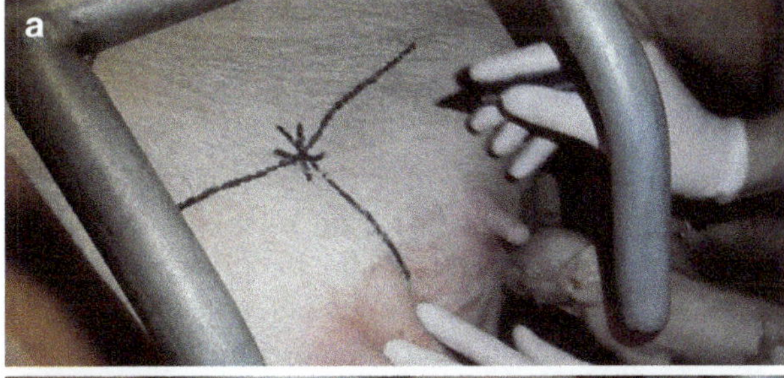
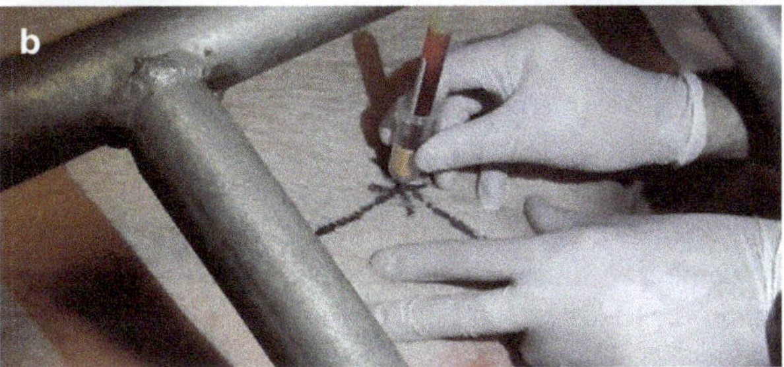

Fig. 3.9 (**a**) Locating the mammary vein (cranial epigastric vein) and (**b**) blood collection in a lactating sow (Photo credit: Claudio Trombani, Breizhpig, France)

2. Clean the general area of the puncture site (at least with water).
3. Disinfection with any disinfecting solution is advised.
4. After positioning the needle just above the mammary glands, a puncture is made perpendicularly between the two mammary glands.
5. When using a syringe, apply a small vacuum shortly after puncturing the skin by gently pulling on the plunger of the syringe. Meanwhile, when using a vacuum tube, connect it immediately via the needle holder after the needle punctures the skin.
6. Push the needle deeper and immediately stop once blood is freely flowing into the tube or syringe (Fig. 3.9b).
7. Once the required volume of blood is collected, stop pulling on the plunger of the syringe or immediately detach the vacuum tube.
8. Carefully remove the needle and immediately apply pressure in a circular motion for around 30 sec on the puncture site after it exits the skin. Clean blood on skin, if any.

3.8.3 Post-procedure Observation

Check the puncture site for any bleeding or bruising. If there is excessive bleeding, apply pressure to the puncture site for a couple of minutes. When returned to the pen, the pig should be moving correctly with a proper gait. If the procedure is performed on anaesthetised pigs, they should only be returned to the pen once they have fully regained consciousness, appear healthy and are moving well.

3.9   Blood Collection from the Ear Vein

3.9.1   Puncture Site and Pig Handling

The ear veins are some of the most superficial veins in pigs. They can easily be dilated using a tourniquet or manual pressure on the vessel. The disadvantage is that these veins are relatively small and collapse easily when blood is withdrawn, and pigs generally dislike having their ears touched. Consequently, a maximum of 3 ml of blood can be taken from the ear veins (Parasuraman et al. 2010; Marchant-Forde and Herskin 2018). Nevertheless, this collection method has proven suitable for single and repeated blood sampling from pigs (Swindle et al. 1992). Anaesthesia is preferred for this procedure. Anaesthetised pigs can be held on the floor, either on their side or with their belly to the ground. Awake pigs can be restrained using a snout rope, as they will shake their head if awake. Although pigs dislike having their ears touched, they can be trained to stand still for venipuncture using positive reinforcement techniques (see Sect. 3.5.2). The use of local anaesthetic gel is also highly recommended.

3.9.2   Description of the Procedure

1. Cleaning and disinfection of the skin of the backside of the ear with any disinfecting solution is advised.
2. If an anaesthetic gel is applied to the skin to be punctured, wait for at least 15 min before doing any restraining and collection (including the tourniquet).
3. Place a tourniquet at the base of the ear (Fig. 3.10).
4. Locate the ear veins. Choose the biggest one possible, as smaller veins can collapse immediately upon using a vacuum, making the blood collection impossible. Warming the ears helps dilate the veins. This can be done by scrubbing the ear with a gauze soaked with warm water.

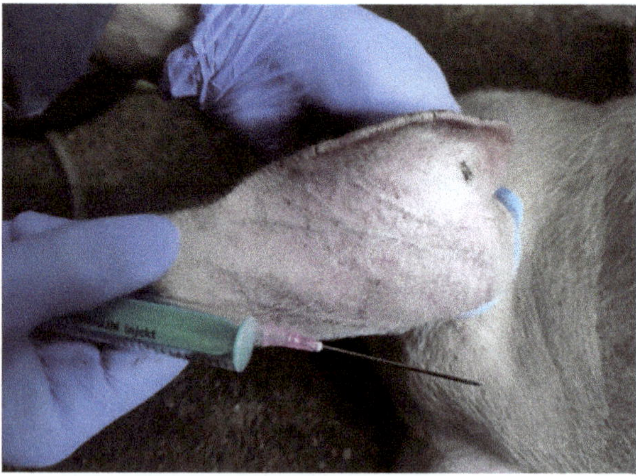

Fig. 3.10 Locating the vein used for the ear venipuncture (Photo credit: Catherine Ollagnier, Agroscope)

5. Puncture is made only when you can palpate and see the vein under the skin. It should be made parallel to the ear vein with the needle pointing towards the ear base, at an angle where the needle is as close to the skin as possible.
6. Remove the tourniquet.
7. When using a syringe, apply the smallest possible pressure. Pulling on the plunger of the syringe during collection should be done slowly and very carefully. Meanwhile, only attach a vacuum tube (the smallest possible) after the needle is well placed in the puncture site.
8. Once the required volume of blood is collected, stop pulling on the plunger of the syringe or immediately detach the vacuum tube.
9. Carefully remove the needle and immediately apply pressure in a circular motion for around 30 sec on the puncture site after it exits the skin. Clean blood on skin, if any.

3.9.3 Post-procedure Observation

Check the puncture site for any bleeding or bruising. If there is excessive bleeding, apply pressure to the puncture site for a couple of minutes. When returned to the pen, the pig should be moving correctly with a proper gait. If the procedure was performed on anaesthetised pigs, they should only be returned to the pen once they have fully regained consciousness, appear healthy and are moving well. If a snout rope was used, check for possible injury to the snout or broken teeth.

3.10 Blood Collection by Catheter from Pigs over 20 Kg

3.10.1 Pig Handling

The following sampling method is described for indwelling catheters with a one-way stopcock installed at the end. The catheter and stopcock should always be kept in place using a medical bandage around the pig's body. This keeps the catheter clean and protects it from damage, such as chewing or grinding against the wall or cage (Fig. 3.11). As the catheter provides a route between the contaminated external environment and the sterile internal environment, protecting it reduces the risk of infection and other health problems for the pig, such as clots, air bubbles and septicaemia. In addition, this method has been tested and verified in individually housed pigs. To minimise stress during blood collection, it is strongly recommended to familiarise the pig with humans. One example of a socialisation training protocol using positive reinforcement training (PRT) can be found in Rydén et al. (2019).

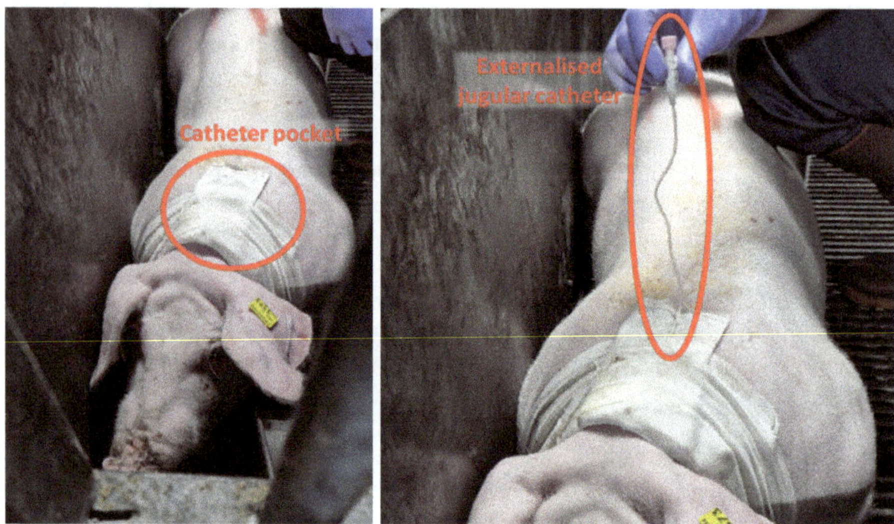

Fig. 3.11 Medical bandage around the pig's neck that protects the catheter (Photo credit: Francis A. Eugenio, INRAE)

3.10.2 Prerequisites and Preparation

Externalised catheters require consistent flushing with saline and regular infection checks at the point of externalisation. Although it should be painless for the animal, collecting blood from the catheter requires a high level of care and hygiene of the catheter and stopcock during each collection. Therefore, in addition to animal handling, sufficient training in catheter handling is required.

The pig should also be trained to tolerate human contact and pouch manipulation. This will make activities much more convenient for personnel and less stressful for pigs. Provided the length of the externalised catheter is sufficient, the pig can move freely during the collection. However, catheterisation usually requires that pigs are isolated from other group members, which can also be stressful for them.

3.10.3 Description of the Procedure

1. Always observe proper hygiene: Before, during and after collection.
2. The tip of the catheter should be carefully removed from the bandaged area that keeps the catheter end clean and protected from damage.
3. Before drawing blood, confirm the catheter is correctly locked (stopcock or cap) to prevent blood from coming out of the catheter.
4. Once done, clamp the catheter before removing the cap and attach a syringe to the end of the catheter. If the catheter is equipped with a stopcock, the position should be changed to allow blood to be drawn.

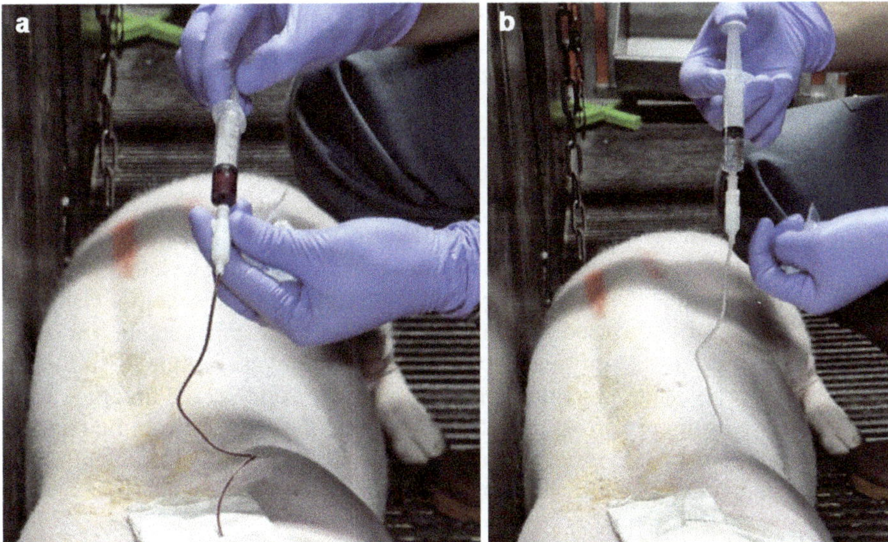

Fig. 3.12 (**a**) Collection of blood with a single-use syringe and (**b**) Flushing of the catheter using sterile saline (Photo credit: Francis A. Eugenio, INRAE)

5. The plunger of the syringe should be drawn back gently until a small amount of blood (< 1 mL) enters the syringe.
6. Turn the stopcock to close or clamp the catheter and then remove the syringe and discard its contents, as it contains a mix of blood and saline solution.
7. Quickly attach a fresh syringe to the catheter, turn the stopcock to open the catheter or remove the clamp and gently draw the syringe plunger back to collect the required volume of blood (Fig. 3.12).
8. Close the stopcock or clamp the catheter and remove the syringe.
9. If the blood sampling is time sensitive, empty the blood from the syringe into the requisite blood collection tubes, mix the tubes gently but thoroughly, and proceed to flushing the catheter. Discard the syringe.
10. If the blood sampling is not time sensitive, set the syringe with the collected blood aside and proceed to flushing the catheter with the saline solution. The blood can be emptied into the requisite tubes after the catheter has been flushed. Discard the syringe.
11. To flush the catheter, take the syringe containing the pre-prepared saline solution and attach it to the catheter.
12. Open the stopcock or remove the clamp and flush the catheter by gently pushing all the saline solution into the catheter. Once the syringe is empty, close the catheter (stopcock and replace the catheter cap). Discard the syringe.
13. After collection, the tip of the catheter can be rinsed with alcohol (Note: this step is not recommended before collecting blood because alcohol can haemolyse collected blood). Return the catheter to its holding place.

3.10.4 Post-procedure Observation

After collection, the pig should appear healthy and move well. Monitoring the pig's body temperature (to assess the risk of infection) and level of activity can serve as a general health check. There should be no visible irritation to the area where the catheter is externalised, nor to the bandage securing the catheter. Ensure that the catheter and stopcock are secured inside the holding pouch.

Ensure that there are no air bubbles in the catheter, as these can be lethal if the catheter is in an artery. Coagulated blood or a fibrin clot can also be lethal, so use an anticoagulant mixed with saline solution to fill the catheter to prevent this. Catheters need to be flushed and rinsed with an anticoagulant and saline solution mixture every two or three days to prevent clotting. The maximum usage time for a catheter depends on the pig's age. In rapidly growing pigs, for example, the tip of the catheter will move as the pig grows, eventually stopping the catheter from functioning when it reaches a smaller vessel. On average, a catheter remains functional for no more than one month in a growing pig and several months in an adult pig.

3.11 Compliance with the 3R Principles

When planning for an experiment, considerations should be made to look for alternative, more refined methods to alleviate stress and pain to the animal during collection. Currently, some parameters (i.e. cortisol (ATOL_0002287)) can be measured in other matrices (e.g. saliva, hair) instead of blood, for which the sampling and restraining methods may be less invasive. Minimally invasive sensors can also be considered to monitor some blood parameters (e.g. glucose, ATOL_0000097). Laboratory techniques are also now available to analyse microvolumes of blood, allowing for safer and less invasive blood collection for the pigs. Collection techniques like pricking the pig's ears using a lancet to collect just a drop of blood may also be considered by future experiments. As usual in animal experimentation, the number of animals required should be based on a power calculation. In addition, the frequency of blood sampling should always be kept to the minimum necessary.

The method and duration of restraining is critical for the extent of stress and/or pain for the animal. Restraining time should be monitored and restricted to the minimum, as recommended in the above sections. Moreover, the duration of restraining has been shown to influence the concentrations of some analytes in blood (Merlot et al. 2011). Some restraining methods are better tolerated by the pig (e.g. pig sling), which should be considered as refinement, however, visualisation and palpation of the puncture site (and the vein) may be more challenging in such a device. Blood sampling can also be refined by socialising and habituating the pigs to humans and the environment where sampling will be performed, as well as to specific handling procedures (e.g. being trained to stay calm in the pig sling). This is important, especially if sampling is repeated. Such training has been very successful in reducing stress levels of the pigs and facilitating the work for the personnel performing the sampling (Fiderer et al. 2024; Rydén et al. 2019; Yang et al. 2021). Implementing

PRT also provides the animals with cognitive enrichment, which further improves their welfare (Sørensen 2010).

3.12 Conclusion

Collecting blood from a pig is not an easy procedure. They do not have many superficial or easily accessible blood vessels. They can also be fearful and are strong animals, so restraining them during blood collection can be dangerous for untrained personnel. In fact, pigs must be immobilised during the collection process. Consequently, blood collection is quite stressful for the pigs, especially if repeated collections are required. Therefore, this activity requires a high level of skill, extensive training and regular practice.

When an experimental protocol requires an analyte to be measured, the possibility of measuring the analyte in a different matrix that is easier to collect than blood (e.g. saliva) should be considered. If blood is required, alternatives to traditional catheterisation or venipuncture should be investigated to minimise the impact of the sampling procedure on the animal. The number and frequency of blood samples should then be kept to a minimum. Finally, the choice of blood collection method and site should be based on the required blood volume and sampling frequency. In practice, however, not all procedures have the same consequences for the animal. Some involve short-term pain (e.g. when the animal is restrained with a snout rope), while others involve medium-term stress (e.g. when individual housing is imposed). These aspects must also be considered when choosing the most appropriate blood collection method.

For all the different sampling methods described, compliance with the 3R-concept should be applied by socialising and training the pigs to reduce their stress levels during sampling. Positive reinforcement training of experimental pigs is proven to be highly successful in facilitating blood sampling, making the procedure less stressful for both animals and humans.

Acknowledgements The authors would like to thank Dr. Daniel Colombus (Saskatchewan University, CA) for reviewing and improving this document.

References

Bollen PJ, Hansen AK, Alstrup AKO (2010) The laboratory swine. CRC Press, Boca Raton. https://doi.org/10.1201/9780849376283

Diehl KH, Hull R, Morton D et al (2001) A good practice guide to the administration of substances and removal of blood, including routes and volumes. J Appl Toxicol 21:15–23

Ferchaud S, Grivault D, Le Floc'h N et al (2019) Prélèvement de sang à la veine saphène chez le verrat lors des collectes de semence. In: Colloque du RMT "Bien-être animal", Strasbourg

Fiderer D, Thoene-Reineke C, Wiegard M (2024) Clicker training in Minipigs to reduce stress during blood collection—an example of applied refinement. Animals 14(19):2819. https://doi.org/10.3390/ani14192819

Marchant-Forde JN, Herskin MS (2018) 16 - pigs as laboratory animals. In: Špinka M (ed) Advances in pig welfare. Woodhead Publishing, pp 445–475

Merlot E, Mounier AM, Prunier A (2011) Endocrine response of gilts to various common stressors: a comparison of indicators and methods of analysis. Physiol Behav 102:259–265

Parasuraman S, Raveendran R, Kesavan R (2010) Blood sample collection in small laboratory animals. J Pharmacol Pharmacother 1:87–93

Rydén A, Manell E, Biglarnia A et al (2019) Nursing and training of pigs used in renal transplantation studies. Lab Anim 54(5):469–478

Scollo A, Bresciani C, Romano G et al (2019) A novel blood-sampling technique in lactating sows: the mammary vein route. Vet J 254:105397. https://doi.org/10.1016/j.tvjl.2019.105397

Straw BE, Zimmerman JJ, D'Allaire S et al (2006) Diseases of swine, 9th edn. Wiley, Ames

Swindle MM, Moody D, Phillips L (1992) Swine as models in biomedical research. Iowa State University Press

Sørensen DB (2010) Never wrestle with a pig. Lab Anim 44:159–161. https://doi.org/10.1258/la.2010.009100

Yang H, Galang KG, Gallegos A et al (2021) Sling training with positive reinforcement to facilitate porcine wound studies. JID Innov 1(2):100016. https://doi.org/10.1016/j.xjidi.2021.100016

Wolfensohn S, Lloyd M (2003) Conduct of minor procedures. In: Wolfensohn S, Lloyd M (eds) Handbook of Laboratory Animal Management and Welfare. Blackwell Publishing Ltd, pp 150–151

Collection of Saliva Samples for Non-Invasive Hormone Measurement

4

Liza R. Moscovice ⓘ and Ulrike Gimsa

Abstract

The advancement in minimally invasive methods to measure analytes in animals has important implications in the applied animal sciences. Techniques that minimise pain and distress are not only advantageous for promoting positive welfare, but also for facilitating repeated measurement of analytes that can be related to behavioural, cognitive and/or emotional responses. While the measurement of hormones in pigs has typically relied on blood sampling, saliva offers an alternative that is easy to collect voluntarily from pigs in their home environments. Many hormones can be readily measured in saliva in pigs, but further refinement and standardisation of best practices for sample collection are greatly needed. Here, we present a detailed protocol for the collection of saliva for the measurement of hormones in pigs. We give recommendations for: (1) Optimal time frame and conditions for capturing changes in peripheral hormones related to specific stimuli of interest in pigs, (2) Materials for sample collection, (3) Step-by-step procedure for collecting samples from individual pigs in their home environments, and (4) Initial laboratory processing steps prior to hormone measurement. We provide detailed descriptions of procedures and focus throughout on assuring high-quality samples for hormone measurement, while minimising any potential stress to pigs related to sample collection.

Keywords

Pigs · Swabs · Habituation · Sample collection · Sample processing

Supplementary Information: The online version contains supplementary material available at https://doi.org/10.1007/978-3-032-24899-2_4.

L. R. Moscovice (✉) · U. Gimsa
Psychophysiology Working Group, Research Institute for Farm Animal Biology (FBN), Dummerstorf, Germany
e-mail: moscovice@fbn-dummerstorf.de

L. R. Moscovice et al. (eds.), *Standard Operating Procedures for Better Pig Research (SOPig)*, https://doi.org/10.1007/978-3-032-24899-2_4

4.1   Introduction

There is increasing interest in developing minimally invasive and non-invasive methods to measure peripheral concentrations of hormones that have traditionally been measured more invasively in plasma (reviewed in Crockford et al. 2014). The use of less-invasive methods facilitates repeated sampling of individuals while ensuring minimal interference with their ongoing behaviour and health (ATOL_0000928). Such an approach is especially critical for research that seeks to investigate affective states, which could be influenced by the sampling method (for example, see Marcet Rius et al. 2018). Developing less invasive methods to measure hormones is also consistent with the refinement of techniques under the 3R principles, which include the development of methods to minimise pain (ATOL_0000863) and distress in order to improve animal welfare (ATOL_0000765) (MacArthur Clark 2018). Among the alternatives to plasma sampling, saliva is increasingly utilised, in part because it is easy to collect samples from individuals at specific time points within their home environment, without requiring restraint. However, recent studies reveal that the method of collection (e.g. passive drool vs. spitting vs. swab-based sampling) as well as the collection material used (e.g. cotton vs. synthetic swabs) can have important impacts on the measurement of analytes in saliva (Bellagambi et al. 2020). This protocol will ensure best practice methods for the collection of saliva to facilitate the measurement of hormones in pigs.

4.2   Goal and Scope of the Procedure

The goals of this protocol are to (1) describe best practice methods for the collection of saliva samples from pigs (beginning from ~3 weeks of age to adulthood), for the purposes of measuring hormones in saliva, (2) identify optimal sampling times to capture changes in peripheral hormones related to specific stimuli of interest in pigs and (3) standardise methods and timing of sampling to avoid unintended impacts of variation in husbandry conditions, time of day, method of sampling or other extraneous factors on hormone measurements. The method is designed for sampling of individual pigs while they are housed in social groups (no separation of pigs is required for sampling). The amount of saliva collected with this method depends on the subject's age and the duration of the sampling, but typically results in approximately 0.1 mL per minute of sampling (referring to the duration of time that the swab is in the pig's mouth) in 1-month-old animals, and 0.2–0.4 mL per minute sampling in older animals.

4.3   Materials and Equipment

4.3.1   Materials for Collection of Samples

- Synthetic swabs; recommended brand: Salimetrics® SalivaBio Infant Swabs (https://salimetrics.com/collection-method/infant-swab-device/) for pigs up to approximately 3 months of age (Fig. 4.1a), and the larger Children's Swabs (https://salimetrics.com/collection-method/childrens-swab-device/) for mature pigs. Note that other synthetic swabs can also be used. Another popular swab is the Sarstedt Salivette® (https://www.sarstedt.com/en/products/diagnostic/salivasputum/product/51.1534.500/).

- Wooden dowels; recommended dimensions approximately 200 mm x 3 mm, L x Ø (Fig. 4.1b). Different wooden dowels or other materials are also fine. Another recommendation is to use surgical clamps to hold the swabs, rather than securing them on wooden dowels. Please note that any material that is reused for sampling should be cleaned thoroughly with water and dried in between each use. Also, make sure that the material you plan to use is not likely to cause cuts to the animals if they chew on it.
- Laboratory gloves (for handling of swabs).
- Polypropylene round-bottom collection tubes with insert for centrifuging (Fig. 4.1c; recommended brand: Sarstedt 14 mL tubes (No. 55.538, dimensions: 105 mm x 16.8 mm, L x Ø). Disposable syringe cylinders (Covetrus, No: 9003018, volume: 5 mL, dimensions: 75 mm x 15 mm, L x Ø) can be inserted into the tubes to hold the samples. Other brands and tubes can also be used, as long as they are made of a polypropylene material. For example, Salimetrics® also offers collection tubes.
- Styrofoam container with ice.
- Rack to hold collection tubes.
- Watches or clocks to record the time of sampling.

4.3.2 Materials for Processing of Samples

- Centrifuge (ideally with cooling function to 4 ° C).
- Pipettor (300–1000 µL) and pipette tips.
- 1.5 mL Eppendorf tubes (Fig. 4.1c), for aliquoting eluent after centrifuging).
- Rack to hold Eppendorf tubes.
- Permanent lab marker (for writing on tubes). Alternatively, pre-printed tube labels can be used if these are available.
- Freezer (−20 to −80 ° C).

4.4 Prerequisites and Preparation

4.4.1 Identifying Ideal Time Frame and Conditions for Sample Collection

Whenever possible, sample collection should occur at the same general time of day across the different test days and test conditions (e.g. always occurring between 10:00 and 12:00 h or always between 13:00 and 15:00 h, just as examples). Sampling should also generally be avoided for the first hour immediately following daily waking (or whenever lights are first turned on in the facility). This is because cortisol

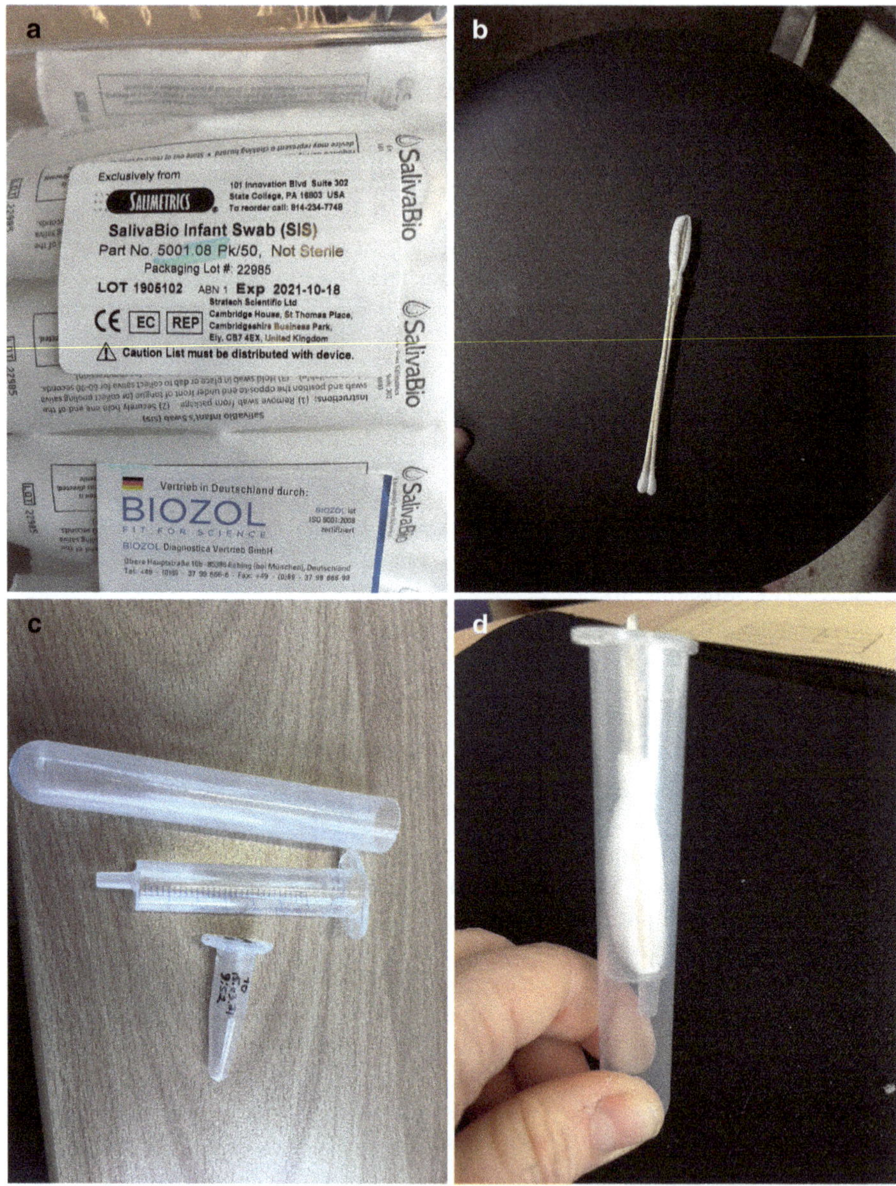

Fig. 4.1 Recommended materials for saliva sample collection. (**a**) Package of Salimetrics SalivaBio Infant Swabs. (**b**) Swab taped to a wooden 'handle'. (**c**) Sample collection tube with insert for centrifuging swab and pre-labelled Eppendorf tube for aliquoting eluent after centrifuging. (**d**) Swab inserted into the collection tube, with the handle broken off prior to centrifuging. Photo credits: Liza R. Moscovice and Arvind Kurup, FBN

(ATOL_0002287) has a natural diurnal rhythm, which peaks within the hour following waking in pigs (Ekkel et al. 1996), and can influence other hormone measurements as well. For similar reasons, when pigs are kept in facilities with controlled lighting and limited natural light, all efforts should be made to standardise the light-dark cycles across sample days, as changes in the light-dark cycle will influence the circadian rhythm (ATOL_0000867), with implications for hormone release, especially in response to stressors (Ruis et al. 1997). Efforts should also be made to avoid sampling subjects soon after eating or drinking, as both may interfere with the measurement of hormones in saliva (McLean et al. 2018). For this reason, sampling should occur outside of set feeding times, and if possible, eating and drinking should be prevented for all pigs beginning 10 min before sampling and continuing until sampling is completed. If this is not possible, monitor the pigs closely during collection and avoid taking a sample from a pig for 10 min after they have eaten or drunk.

Important

The ideal sampling time frame will depend on the research goals. For example, appropriate sampling regimens to address questions about variation in baseline or average hormone concentrations over time, or in different environments, will differ greatly from the ideal sampling regimen to capture short-term changes in hormones related to specific events or stimuli. Even when focusing on capturing short-term changes in hormones, the duration and intensity of the stimulus will influence the ideal timing for sampling to detect changes in hormones. Depending on the specific stimulus, there may be a delay of between 10 and 15 min between changes in peripheral hormone concentrations and detection in saliva (de Jong et al. 2015; Heinrichs et al. 2003). Salivary hormone concentrations typically return to baseline levels between 40 and 60 min following the end of a stimulus (de Jong et al. 2015; Merlot et al. 2011; Ruis et al. 2001). Thus, to capture short-term changes in hormones, it is recommended to collect the following samples whenever possible:

1. Pre-event sample: Collected prior to the stimulus, and when the pigs have not had any major disruptions for the previous 45 min.
2. Event sample: Collected between 15 and 20 min after the onset of the stimulus. Adjustments can be made to fit the specific study design. For example, if the stimulus lasts for 30 min and collection before that would interfere with the stimulus itself, then the event sample can be collected immediately at the end of the stimulus.
3. Post-event sample: Collected between 45 and 60 min after the end of the stimulus.

4.4.2 Preparation of Materials for Sample Collection

First, prepare the wooden 'handle', if using this for collection. This is to facilitate sample collection without putting your fingers near the pig's mouth. We also find that pigs prefer to bite on the swab when it is attached to something hard. Using clear cellophane tape, tape two wooden dowels together around the middle. Two dowels provide more support than one and make it less likely that the pig will break the dowels. Wear lab gloves and then open one end of the swab package to slide a swab out. Fold the swab over the top of the sticks and use tape to attach the swab securely at its base to the sticks (Fig. 4.1b). The goal is to ensure that the swab is secure enough that the pigs are not able to detach the swab from the stick, in which case you could lose the sample. However, you should try to cover as little of the actual swab with tape as possible, so that there is more absorbable substrate for the saliva. Swabs can be prepared a few days in advance and then slid back into the packaging to keep them clean until collection. You will need to prepare one swab plus handle per subject per sample, plus a few extras in case of re-sampling.

It is also helpful to pre-label swab collection tubes (the large tubes with inserts that will be put in the centrifuge) with a short name. These labels are only for the short-term, since the samples will be stored in fully labelled Eppendorf tubes after centrifuging. It may be enough to indicate each pig's ID number and record whether it's sample 1, 2, 3, etc., for that pig on that collection day.

You can also pre-label the Eppendorf tubes, or you can choose to do that on the day you process the samples. For labelling the Eppendorf tubes for long-term storage, you should have a recording system that can easily link each tube to a specific pig's identity, the date and time of sample collection (which can be important control predictors to include in statistical models) and the specific context or condition in which the sample was taken (e.g. pre-event, event, post-event, etc.). This is probably too much information to record directly on the tube, but you should find a system so that enough information is on the tube to make it uniquely identifiable (e.g. pig ID, date and context), with the rest of the required information contained on a separate spreadsheet that is linked to each unique Eppendorf tube label. We recommend writing the pig ID and context on the top of the tube (e.g. "Pig1, pre"), and pig ID, date and context on the side of the tube (e.g. "Pig 1, pre, 21.10.25"). These are just examples and you should adapt the labelling to your own needs. Some repeat information on tubes helps in case of marker fading. Since the exact time of collection will not be known in advance, you can record that on a separate spread sheet that is linked to each sample. Other relevant information about eating or drinking prior to collection, health concerns about the pigs, volume of samples collected, etc., can also be recorded on this separate spreadsheet (refer also to Supplementary Table 4.1).

4.4.3   Habituation of Pigs

Prior to sample collection, habituate subjects to the researchers and procedure (we recommend 3–5 habituation visits, of 30 min to one hour per visit). This will make sample collection on test days much faster and reduce the chances of causing stress to the pigs during sampling. You should do the habituation similarly to the conditions under which you will collect the samples. If you plan to enter the stall to collect the samples, you should also enter the stall during the habituation. If you plan to collect the samples from outside the stall, you should probably remain outside the stall during the habituation. Either way, you should first let pigs get comfortable with your presence, and let them approach you voluntarily. When pigs approach, you can attempt to interact with them, for example, by letting them sniff your hands, stroking their heads and bodies if they allow this and giving them something to chew on (i.e. a rope or other toy). Then present pigs with a few practice swabs and give them a chance to explore and manipulate the swabs. If you are in the stall, it may help to crouch down and first hold the swabs close to the floor, so that pigs can explore and sniff them. Then try to gently insert the swabs in their mouths, ideally on either side, between the cheek and gums or under the tongue (think about where saliva pools in your own mouth and that is the location you want to aim for in the pigs). Try to keep the swab in contact with the inside of the mouth for between 1 and 2 min in total. If a pig will not continuously chew on or hold the swab in its mouth for the full time, you can pause and then try to interact with the pig again. During the habituation sampling only, you can touch the swab to check how much saliva you are collecting. If the swab is wet to the touch, this is a good sign that you have probably collected enough saliva (e.g. between 0.2 and 0.5 mL). If the swab is still pretty dry, you have probably obtained less than 0.2 mL of saliva and should continue to encourage the pig to hold the swab in its mouth. If you find that the pigs are not willing to accept the swab, you can increase their motivation by first dipping the swab in juice, apple sauce, or another sweet-tasting substance and then offering it to them. This is possible during the habituation trials since samples will not be analysed. Note that as stated in Sect. 4.4.1, no food or liquid should be provided before or during the collection of actual research samples.

4.4.4   Identifying Individuals

This protocol assumes that you want to measure hormone concentrations in known individuals. If your goals are rather to determine average concentrations of hormones within a group, it may be sufficient to sample from several pigs within a group without identifying individuals. However, when it is necessary to know the identity of each subject for hormone analyses, we recommend labelling of subjects with easy-to-read identification numbers using a non-invasive method, such as a non-toxic dye, which can be applied to each pig's back to facilitate identification for sample collection. Back numbers should be linked to long-term identification numbers (e.g. ear tags) and data (e.g. sex, genetic background) of each pig. Back numbers should be applied 1–2 days before the first sampling and updated when fading. Pigs can also be habituated to receive back numbers without being restrained (e.g. by providing a preferred food reward and back numbering while the pigs eat the reward). Do not back number pigs on sample collection days, to avoid any interference from back labelling with hormone measurements.

4.4.5 Keeping Samples at Appropriate Temperatures

Whenever possible, samples should be kept cold during collection by keeping them on ice. They should also be processed using a centrifuge that cools to $4\,^{\circ}$ C. These steps help to reduce potential degradation of hormones that can occur in samples kept at higher or more fluctuating temperatures. In case you do not have access to ice or to a centrifuge with a cooling function at your pig facility, you can also collect and process the samples at room temperature- this will have little effect on some hormones, such as cortisol, which are relatively stable at different temperatures, but it could affect measurement of other hormones such as oxytocin (ATOL_0006118), which can degrade faster at higher temperatures. It is therefore important to tailor the sampling procedure to the specific hormones that you are interested in measuring. If you do not have access to a centrifuge at your facility, you should make arrangements for samples to be transported on ice to a facility with a centrifuge as soon as possible following collection (e.g. within one hour), for the additional processing steps. Alternatively, swabs containing saliva can be stored in polypropylene tubes and frozen directly after collection at -20 to $-80\,^{\circ}$ C. Swabs should then be thawed and processed using a centrifuge within 3 months of collection.

4.5 Description of the Procedure

4.5.1 Immediately Prior to Sample Collection

Turn on the centrifuge and set a program for 15 min at $4\,^{\circ}$ C and 4000 x g. It takes approx. 30 min for the centrifuge to cool down to $4\,^{\circ}$ C. Keep the centrifuge on until all sample processing is completed. Prepare ice in a large Styrofoam container to use during sample collection. If you have not already pre-labelled the swab collection tubes, this should be done now (Sect. 4.4.2). Put the labelled swab collection tubes with their inserts in a rack on the ice. Your swabs plus the holding method (e.g. taped to wood dowels, held with surgical clamps or other method) should already be prepared (Sect. 4.4.2).

4.5.2 Sample Collection

Bring the swabs, the Styrofoam container with ice, pens, a watch and a data collection sheet into the animal room. If you will enter the pen to collect the samples, it is recommended to keep all of these materials on a cart outside of the pen that can be reached from inside the pen. If it is possible that the animals have eaten or had a drink recently, then first observe the pigs for ten minutes before beginning collection, to ensure that no eating or drinking occurs immediately before collection. Identify your first subject (ideally using easy-to-see back numbers) and approach the pig. If the exact order of collecting the samples from each subject does not matter, you can wait to see which subject approaches you first, and then they can provide the first sample. Note that if you are trying to collect separate samples from each individual, then once a swab has been in one pig's mouth, it is important to avoid letting any other pigs come in contact with that swab. You can let other pigs sniff your hands and other body parts, but make sure to keep the swab away from other pigs if you have already begun collection on a specific subject. If there are two experimenters collecting samples, it is also sometimes helpful when a second experimenter distracts the other pigs, while one experimenter focuses on collecting a sample from the target subject.

If pigs are scared, it is best to crouch down and remain further away from the pigs, with your hand extended with the swab, and let the pig approach you. However, if you follow the habituation steps (Sect. 4.4.3), this should reduce fear and avoidance in most pigs. Place the swab in the pig's mouth, trying to reach the side of the mouth between the cheek and the gums, where saliva pools (Fig. 4.2a). If possible, try to avoid putting the end of the swab, where it is attached to the handle, or the

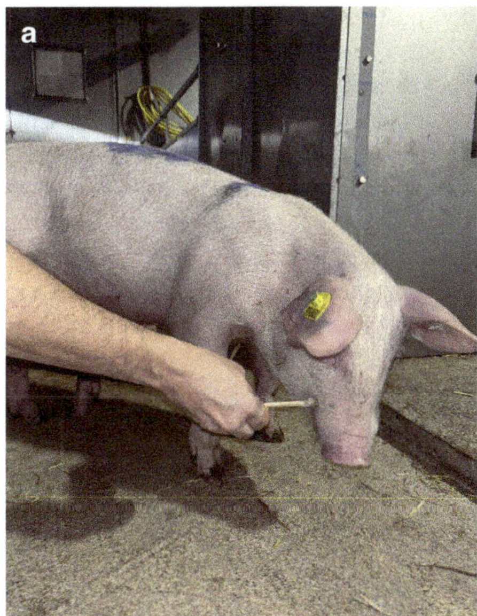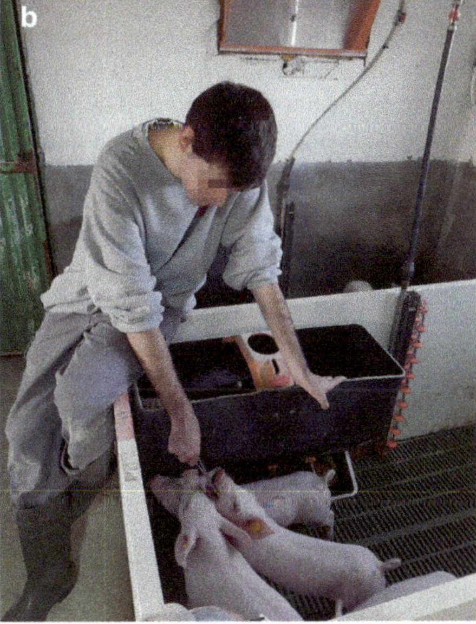

Fig. 4.2 Examples of voluntary saliva sampling from individual pigs while in their home environments. Swabs can be attached to wooden dowels (**a**) or held with surgical clamps (**b**). Photo credits: **a**) Arvind Kurup, FBN, **b**) Berta Baulida, IRTA)

handle itself, in the pig's mouth, so that as much of the saliva as possible collects on the swab itself and not on the taped part, or on the wooden dowels. Some pigs will accept the swabs easily after habituation. Others will be less enthusiastic, so it will take some patience, but all pigs should be able to provide samples without being restrained. Try to keep the swab in contact with the inside of each pig's mouth for 1–2 min (this need not be continuous). Make sure that your subject does not eat or drink during sample collection. When swabs look wet or well chewed-on, they are ready. If a pig detaches the swab from the stick, do not panic! Most likely, the pig will continue to chew on the swab for a few seconds and then spit it out, at which point you can collect and use it as long as it is not contaminated with faeces or saliva from other pigs. If the swab is swallowed or becomes contaminated, then try to collect a second sample. Take precautions to avoid being bitten by avoiding putting individual fingers near the subject's mouth during collection. Rather, make a fist and hold the swab in your fist. This will reduce the risk of being bitten.

Try to finish the sample collection for all subjects within 15–20 min. From our experience, this should be enough time to sample between 5 and 10 pigs per person. Sampling within a short time interval ensures that all pigs are sampled within a relevant time period following the stimulus of interest and also reduces any potential impact from the sampling procedure itself. For larger groups, this may mean having more than one researcher collecting samples from the group at the same time, if possible.

Inspect the swab for any signs of blood (ATOL_0005631) (small red spots on the swab). Dirt is not a big problem, but blood will interfere with hormone measurement. If you suspect that there may be blood on the swab, it should be discarded. Collect a second sample using a new swab and try to avoid contact with the inner cheeks (in case there is a wound there). Inspect the second swab. If it also has blood, discard that sample as well and mark this as a missing sample for that pig. Place the swab (swab side down if attached to a handle) in the insert tube, and place the insert tube in the appropriate pre-labelled collection tube on ice. Record on a separate data sheet the time of collection (to the nearest minute) for each sample. In addition, record further relevant notes about the pig's behaviour, any health concerns, any eating or drinking prior to sampling, etc. Keep samples in the tubes on ice for up to one hour, until collection is finished. If sampling takes more than one hour to complete, it is advised to process samples in smaller batches so that subsets of samples can be processed within an hour of collection.

4.5.3   Post-collection Processing

Within 1 h of collection, bring the samples on ice to a centrifuge. If samples are attached to wooden dowels, break off or remove the tops of the dowels, which may be too tall for the centrifuge. Centrifuge the tubes for 15 min at $4000 \times g$ and 4 °C. If Eppendorf tubes have not been labelled during the prior preparation steps, they should be labelled now (for details refer to Sect. 4.4.2).

After centrifuging, take out and discard the used swab and the insert tube. Aliquot the saliva from the collection tube into the pre-labelled Eppendorf tubes, using a pipettor. The volume of saliva should be between 0.1 and 1 mL. If you have more than 0.5 mL of saliva, it is usually preferable to split the sample into multiple replicates in different tubes, in case samples need to be re-measured, and/or for use in preparing controls for assay validation steps (see Chap. 8). For replicates of the same sample, these can be labelled in the same way as the first sample (e.g. "Pig1, pre"), with an additional code added to the lid to indicate that this is a replicate sample (e.g. "repI", "repII" etc.). Record relevant information about the sample processing (e.g. total volume, number of replicates, any saliva discolouration) on the sample collection sheet (see Supplementary Table 4.1). Note that pink-hued saliva samples may indicate that there is blood in the saliva, so such samples should be analysed with caution (or not at all). Transport samples on ice as soon as possible to a freezer, and store at -20 to $-80\ °C$ until further analysis. For long-term storage (> 1 month), use of a $-80\ °C$ freezer is recommended to minimise potential hormone degradation.

4.6 Compliance with the 3R Principles

The collection of saliva samples for hormone measurement is consistent with the 3R principle of Refinement, which includes the development of methods to minimise pain and distress in order to improve animal welfare (MacArthur Clark 2018). Although blood sampling remains the primary method for hormone measurement in farm animals, it typically requires the removal of animals from their home environment and restraint, both of which may trigger stress responses in pigs (Geverink et al. 2002; Hawkley et al. 2012). Alternatively, animals can be catheterised to facilitate repeated blood sampling without restraint, but this method typically requires that subjects are housed individually without social contact, which can be stressful for pigs. In addition, this method increases the risks of complications from infections and can also cause discomfort to the animals, leading to a classification of catheterisation as causing a 'moderate' degree of pain to subjects under the Animal Welfare Act (Animal Welfare Act 2006). Saliva sampling offers a minimally invasive alternative to blood sampling, which can be achieved without the restraint of animals or removal from social groups. Moreover, saliva sampling does not involve any puncturing of skin (ATOL_0005636) and is associated with minimal pain and stress. This protocol also emphasises the importance of voluntary participation by subjects and habituation of animals to the sampling procedure, which is expected to further reduce the likelihood of any distress associated with the procedure itself.

4.7 Conclusions

We present a standardised protocol for the collection of saliva samples for hormone measurement in individual pigs that is designed to provide high-quality samples for analyses, while reducing any pain or stress to the subjects. Using minimally invasive methods for hormone collection has important advantages for research as well as for welfare purposes, by reducing the chances that hormone measurements reflect responses to the collection procedures themselves, rather than to the stimuli or events of interest. The lack of invasive procedures also means that the sampling method can be performed by a wide range of animal care personnel without requiring veterinary oversight or specialised techniques. By standardising the methods and materials for sampling, as well as the timing of sampling relative to events, this method increases the chances of detecting meaningful changes in hormones related to specific events of interest. The protocol can also be used by collaborating researchers from different facilities/institutes to optimise conditions of sample collection for more direct comparison of hormone concentrations in different pig populations, when relevant. In such cases, all efforts should be made to standardise the methods for laboratory analyses as well, ideally by performing analyses of all samples in a single laboratory.

The disadvantages of this method include the time required to habituate the animals and collect the samples, some safety risks to collectors due to the risk of being bitten during the sampling procedure and minimal safety risks to the pigs, in rare cases in which pigs detach and swallow the swab. However, there is no evidence that pigs suffer adverse consequences from swallowing swabs (L. R. Moscovice, personal observation). When working with large groups of pigs, the duration of time required to collect all of the samples increases, which also increases the chances that hormones may reflect responses to the collection procedure, rather than to events of interest. To reduce this risk, all efforts should be made to use the minimal number of pigs required to detect an effect, e.g. by using power analyses when designing the experiment. In addition, methods for measuring hormones in saliva remain less well-established and standardised compared with methods for hormone measurement from plasma. As a result, care must be taken to run biochemical validations of salivary hormone measurements in the lab (Moscovice et al. 2024, see also Chap. 8). In addition, more work is needed to validate various laboratory methods for measuring hormones in saliva (e.g. MacLean et al. 2019).

Acknowledgements We are grateful to Winfried Otten, Research Institute for Farm Animal Biology (FBN) and Elodie Merlot, National Research Institute for Agriculture, Food and Environment (INRAE), for their valuable feedback on earlier versions of this SOP.

References

Animal Welfare Act in the version published on 18 May 2006 (Federal Law Gazette I p. 1206, 1313), amended by article 4 section 90 of the Act of 7 August 2013 (Federal Law Gazette I p. 3154)

Bellagambi FG, Lomonaco T, Salvo P et al (2020) Saliva sampling: methods and devices. An overview. Trends Anal Chem 124:115781. https://doi.org/10.1016/j.trac.2019.115781

Crockford C, Deschner T, Ziegler TE et al (2014) Endogenous peripheral oxytocin measures can give insight into the dynamics of social relationships: a review. Front Behav Neurosci 8:68. https://doi.org/10.3389/fnbeh.2014.00068

de Jong TR, Menon R, Bludau A et al (2015) Salivary oxytocin concentrations in response to running, sexual self-stimulation, breastfeeding and the Trier social stress test (TSST): an oxytocin challenge study. Psychoneuroendocrinology 62:381–388. https://doi.org/10.1016/j.psyneuen.2015.08.027

Geverink NA, Schouten WGP, Gort G et al (2002) Individual differences in behavioral and physiological responses to restraint stress in pigs. Physiol Behav 77:451–457. https://doi.org/10.1016/S0031-9384(02)00877-6

Hawkley LC, Cole SW, Capitanio JP et al (2012) Effects of social isolation on glucocorticoid regulation in social mammals. Horm Behav 62:314–323. https://doi.org/10.1016/j.yhbeh.2012.05.011

Heinrichs M, Baumgartner T, Kirschbaum C et al (2003) Social support and oxytocin interact to suppress cortisol and subjective responses to psychosocial stress. Biol Psychiatry 54:1389–1398. https://doi.org/10.1016/S0006-3223(03)00465-7

MacArthur Clark J (2018) The 3Rs in research: a contemporary approach to replacement, reduction and refinement. Br J Nutr 120:S1–S7. https://doi.org/10.1017/S0007114517002227

MacLean EL, Gesquiere LR, Gee N et al (2018) Validation of salivary oxytocin and vasopressin as biomarkers in domestic dogs. J Neurosci Methods 293:67–76. https://doi.org/10.1016/j.jneumeth.2017.08.033

MacLean EL, Wilson SR, Martin WL et al (2019) Challenges for measuring oxytocin: the blind men and the elephant? Psychoneuroendocrinology 107:225–231. https://doi.org/10.1016/j.psyneuen.2019.05.018

Marcet Rius M, Cozzi A, Bienboire-Frosini C et al (2018) Selection of putative indicators of positive emotions triggered by object and social play in mini-pigs. Appl Anim Behav Sci 202:13–19. https://doi.org/10.1016/j.applanim.2018.02.002

Merlot E, Mounier AM, Prunier A (2011) Endocrine response of gilts to various common stressors: a comparison of indicators and methods of analysis. Physiol Behav 102(3–4):259–265. https://doi.org/10.1016/j.physbeh.2010.11.009

Moscovice LR, Sobczak B, Niittynen T et al (2024) Changes in salivary oxytocin in response to biologically-relevant events in farm animals: method optimization and usefulness as a biomarker. Front Physiol 15:1370557. https://doi.org/10.3389/fphys.2024.1370557

Ruis MAW, De Groot J, Te Brake JHA et al (2001) Behavioural and physiological consequences of acute social defeat in growing gilts: effects of the social environment. Appl Anim Behav Sci 70:201–225. https://doi.org/10.1016/S0168-1591(00)00150-7

Ruis MAW, Te Brake JHA, Engel B et al (1997) The circadian rhythm of salivary cortisol in growing pigs: effects of age, gender, and stress. Physiol Behav 62:623–630. https://doi.org/10.1016/S0031-9384(97)00177-7

Collection of Urine Samples Using Metabolic Cages or Manually by Spontaneous Urination

5

Miriama Sciascia, Andreas Vernunft, Virág Ács, Maria Eskildsen, and Marion Girard ⓘ

Abstract

The development of minimally and non-invasive sampling techniques to measure current "gold-standard" biomarkers and assess the potential of a new generation of biomarkers is central to future-proofing animal research. Urine can be sampled non-invasively and is easy to voluntarily collect from pigs in confined (e.g. metabolic cages) or open (group or single pens, outdoors) environments. Urine has been utilised in a wide range of animal studies to measure a plethora of parameters, including untargeted and targeted metabolites, clinical chemistry parameters and hormones. Here we present protocols for the collection of urine from pigs housed in metabolic cages, as well as from unrestrained pigs. The metabolic cage method is ideal for continuous sampling, whereas the manual collection by spontaneous urination is ideal for spot sampling in open environments. This SOP provides recommendations for the following: 1) materials for sample collection, 2) step-by-step procedures for collecting samples from pigs and 3) initial urine sample processing steps. The sampling techniques described in this SOP provide reliable methods for the collection and initial processing of samples for the measurement of a range of analytes in pig urine, thereby promoting the development of minimally invasive approaches to measuring biomarkers in animal research.

M. Sciascia (✉) · A. Vernunft
Research Institute for Farm Animal Biology (FBN), Dummerstorf, Germany
e-mail: sciascia@fbn-dummerstorf.de

V. Ács
Department of Farm Animal Nutrition, Hungarian University of Agricultural and Life Sciences (MATE), Kaposvár, Hungary

M. Eskildsen
Department of Animal and Veterinary Sciences, Aarhus University, Tjele, Denmark

M. Girard
Swine Research Group, Agroscope, Posieux, Switzerland

© The Author(s) 2026

L. R. Moscovice et al. (eds.), *Standard Operating Procedures for Better Pig Research (SOPig)*, https://doi.org/10.1007/978-3-032-24899-2_5

Keywords

Pigs · Sample collection · Sample processing

5.1 Introduction

Urine is an important biological sample that has been used in biomedical, veterinary and nutritional research, to investigate a wide variety of research questions (Czech et al. 2022; Kim et al. 2020; Nixon et al. 2020; Tkaczyk et al. 2021). In pigs, urine has been used to:

(1) Assess kidney function. It can be used to measure the renal clearance of certain substances, such as creatinine, urea and insulin (Dhondt et al. 2020; Jia et al. 2010).
(2) Monitor metabolism to study the effects of a treatment or a diet (e.g. protein-rich diets, medications or probiotics), the nitrogen balance (ATOL_0005338) or detect metabolic imbalances (Berghaus et al. 2023).
(3) Study pharmacokinetics to quantify the excretion of substances and their metabolites, assess their bioavailability and study the half-life of a compound (Dhondt et al. 2020).
(4) Measure specific biomarkers and monitor animal health (ATOL_0000928) to diagnose urinary tract infections, track the progression of certain metabolic or inflammatory diseases and detect the presence of abnormal or pathological molecules (Svoboda et al. 2024; Sachse et al. 2016).

There are several methods for collecting urine, including cystocentesis, bladder catheterisation, metabolic cages (EOL_0002023) or manual urine collection by spontaneous urination (ATOL_0000802). The choice of method depends on the type of analysis to be performed, with the caveat that all methods differ in their degree of invasiveness. Cystocentesis and bladder catheterisation are the ideal methods for collecting sterile samples for bacteriological analysis. However, these are highly invasive procedures that carry a risk of infection and require technical expertise. Urine can also be collected using non- or minimally invasive methods, making it a potential candidate to replace more invasive sampling techniques, if and when appropriate urinary biomarkers are found. The two methods described here, metabolic cage and manual urine collection by spontaneous urination, both allow for non-invasive urine sampling.

Metabolic cages are the "gold standard" method when sequential urine samples are required or when a large volume of urine needs to be collected over a particular period. For example, during metabolic and nitrogen balance studies, it is essential to know how much urine is produced over several consecutive days. Unlike faeces, which can be collected as a spot sample by incorporating indigestible markers into the feed (Chap. 6), there are currently no markers in urine to measure total output. Metabolic cages also require significant financial investment, as experiments involving these cages necessitate rooms that are large enough to accommodate the equipment and provide sufficient workspace for staff. Additionally, these rooms must be equipped with systems to control temperature, humidity and water supply. Facilities must also allow for daily cleaning of both rooms and cages, as well as include appropriate waste disposal systems. In addition, their use in the EU is heavily restricted due to the confined nature of their setup, which affects pigs' natural behaviours and can place unnecessary stress (ATOL_0002301) on them. There is currently no age limit on the use of metabolic cages in pigs; studies have been conducted in newborn piglets and in sows over 4 years old. In contrast to metabolic cages, manual urine collection by spontaneous urination can be conducted while pigs remain unrestrained in their normal housing environment. However, the method requires staff to remain in the housing environment and wait for pigs to urinate. This makes manual urine collection by spontaneous urination a time- and personal-intensive method.

To ensure the reliability of urine-based measurements, minimizing sample contamination is critical. This includes avoiding faecal contamination and environmental debris. Proper handling and collection protocols, such as using clean sampling equipment and promptly transferring samples to appropriate storage conditions, are also important. Additionally, reducing stress, which can influence urinary biomarkers, by habituating animals to the sampling environment (metabolic cages) or to the presence and proximity of human collectors (manual urine collection by spontaneous urination), helps to ensure that the data reflect true physiological conditions rather than stress-induced changes.

5.2   Goal and Scope of the Procedure

The goal of this SOP is to provide detailed information on the collection, processing and storage of urine samples collected from pigs under the following conditions: (1) individual housing from the post-weaning period onwards in metabolic cages (Sect. 5.3) and (2) individual or group-housing, in pigs of any age, using the manual urine collection by spontaneous urination method (Sect. 5.4).

These guidelines on urine collection have been designed to (1) reduce sample contamination (i.e. from faeces or feed), (2) be easy and convenient to conduct, (3) enable reproducibility, (4) habituate pigs to urine collection, and (5) minimise the stress, pain (ATOL_0000863) or discomfort experienced by the animal, particularly when using metabolic cages.

5.3   Urine Collection from Pigs Housed in Metabolic Cages

5.3.1   Materials and Equipment

5.3.1.1 Metabolic Cage

- Should be adjustable in width and length according to the size of the pig (Fig. 5.1a-c).
- Should be height-adjustable for ease of use by the people working with the pigs.

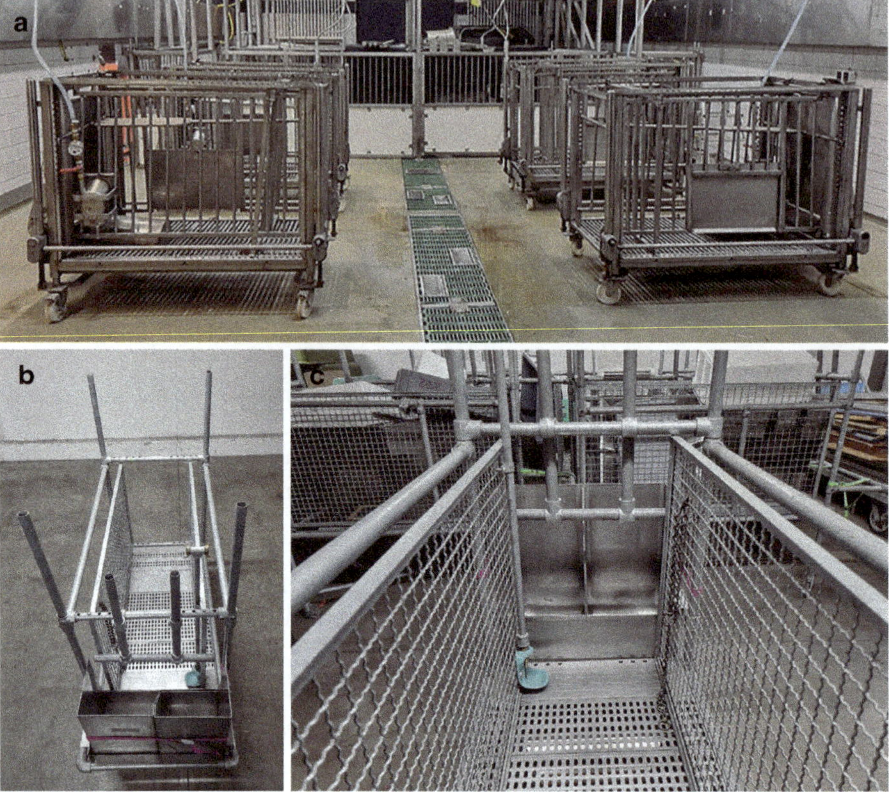

Fig. 5.1 Adjustable metabolic cages used at (**a**) FBN (Credit: Mariagrazia Cavalleri, FBN) and (**b**) Agroscope, with an example of (**c**) a feeding and water system (Credit: Marion Girard, Agroscope)

- Should have a catchment tray or a gradual slope design for catching the urine samples.
- Filter(s) need to be installed at the urine collection point of the floor and/or tray to prevent faecal and feed matter from entering the urine collection vessel (Fig. 5.2a-c).

5.3.1.2 Urine Collection Containers

- The size of the collection container (~1–4 litres) to be used will depend on:
- The size of the pig (bigger pigs produce more urine). A weaner pig (10–15 kg) produces 0.5–1.5 litres per day, whilst a growing-finishing pig (50–100 kg) will produce between 1.5 and 8 litres per day.
- The sampling time duration. In pigs between 20 and 40 kg, a 1 litre bucket is sufficient to collect urine over 3- to 12-h time periods with no preservatives or pH regulators, for the measurement of Carbon-13 (^{13}C) (Cavalleri et al. 2025), mannitol and lactulose. In certain metabolic trials, it is important to know the

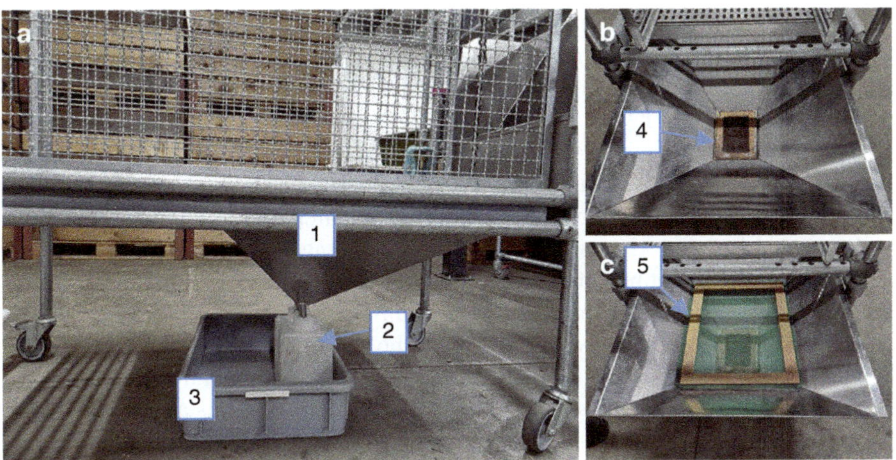

Fig. 5.2 Example of a urine collection system (**a**) [1] Stainless-steel collection tray, with [2] Urine collection container (~2 L) and [3] Container that can be filled with ice to keep the urine sample cold, and top-down views of the stainless-steel collection tray (**b**, **c**) with [4] Secondary 0.5 mm filter, and [5] Primary 2 mm filter (Credit: Marion Girard, Agroscope)

volume of urine produced over 24 h in order to determine the amount of a compound excreted in the urine. For the analysis of nitrogen excreted over 24 h in nitrogen balance studies, urine is collected into two 1.5-litre containers (one acidified and one non-acidified) using a collection tray with two outlets. The fill level of these containers is checked once or twice during the day. If containers are found to be 2/3 full, they must be replaced with new, clean and dry containers.

The intake of the collection container (where the urine enters) should be designed to reduce exposure to the external environment and thus reduce any potential environmental contamination.

- We recommend using 1 litre plastic buckets covered with a stainless-steel round sieve (Rotilabo, Product ID: 8099.1, Roth Labs) that has a 1.5 mm mesh to catch any large environmental particles and acts as a secondary filter for the urine samples. This design has been used for the measurement of ^{13}C enrichment (Cavalleri et al. 2025) and concentrations of lactulose and mannitol (markers of intestinal permeability).
- An alternative option is to use a container with a narrow opening at the top. You should ensure that the end of the collection tray fits properly into the opening of the container, as shown in Fig. 5.2a.

At the end of each experimental period, containers should be cleaned in a dishwasher to ensure thorough cleaning. If the containers do not need to be sterilised, rinsing with hot soapy water and drying may be sufficient. The frequency with which the urine collection containers should be cleaned depends on the trial design and what needs to be measured.

- An example is a study conducted over a 36-h experimental period to measure the ^{13}C abundance in urine, using ^{13}C stable isotope labelled lactose-ureide (Cavalleri et al. 2025). Samples were collected over a 12-h period at 3-hr intervals (0–3, 3–6, 6–9 and 9–12 hr) prior to lactose-^{13}C-ureide administration and from 12 to 24 h post-lactose-^{13}C-ureide administration. After each sampling time-point, the urine collection buckets and stainless-steel round sieves were washed with hot soapy water to remove any residual ^{13}C, immediately before the next urine collection period. As ^{13}C was the only parameter being measured, the urine collection buckets did not need to be sterile and could be reused.
- As another example, if the urine of one pig needs to be collected over several consecutive days, it is advisable to wash the containers with hot water and dry them between each day of collection and to use the same containers for the same animal each day. In the case of nitrogen balance studies, it is advisable to have two containers (one acidified and one non-acidified) to place under the collection tray and two spare containers (one acidified and one non-acidified) if urine production is abundant.

5.3.1.3 Other Materials and Equipment for Urine Sample Collection in Metabolic Cages

- Water-resistant felt-tip pen.
- Lab gloves (latex or nitrile).
- Scale, to weigh the urine collected.
- Styrofoam container with ice to place the bucket/container in; this keeps the collected urine cold and slows down bacterial degradation of target metabolites (Fig. 5.2a).
  - The size of the Styrofoam container and amount of ice to be used will depend on many factors. Some examples are as follows: the length of time urine is collected (longer duration requires a larger container and more ice) and environment (temperature-controlled vs. non-controlled stalls). We suggest the use of thermometers placed in the ice next to the collection container to ensure the temperature is maintained below 4 °C.
  - Ensure that the container can always be easily removed from under the collection tray, even when it is full and submerged in ice. This is especially important when a new container needs to be used.
- Urine preservatives may need to be added for analysis of specific metabolites.
  - For example, to measure nitrogen, it is recommended to acidify urine with sulfuric acid to prevent microbial growth. We recommend adding 25 g of sulfuric acid at 3 M to 1.5-liter collection containers.

5.3.1.4 Other Materials and Equipment for Urine Sub-Sample Preparation in Metabolic Cages

Urine is typically collected in buckets/containers (~1–4 litres), and the large volume produced requires sub-sampling into smaller aliquots, using the following equipment:

- Styrofoam container with ice to transport the sub-samples.
- Lab gloves (latex or nitrile).
- Pipettes and tips (volumes will depend on how much urine is being sub-sampled), to prepare the sub-samples.
- Eppendorf tubes or 1-litre containers for the sub-samples (if needed).
- Some studies require a small aliquot, while others require an aliquot representative of the entire collection period. One example would be collecting 2% of the total daily urine excretion over several days. In this case, larger containers (0.5–1 litre, one per pig) are needed.
- Centrifuge, with 4 °C cooling.
- Freezer (-20 °C or -80 °C), to store the sub-samples until further analysis (if needed).

5.3.2 Prerequisites and Preparation

5.3.2.1 Metabolic Cage Setup and Study Design

Metabolic cages should be in a room where multiple cages can be set up, so that the pigs can hear, see and smell each other. One design is a six-cage setup, with two rows of three cages set across from each other (Cavalleri et al. 2025). The room should also have adequate ventilation and climate control so that the pigs are kept under temperature and humidity conditions according to their age and development stage. For pigs between 20 and 40 kg, we recommend a room temperature of 22 °C and a humidity of 40–60%. In addition, the room should have lighting to control day/night settings. Once the appropriate metabolic cages have been selected, they should be cleaned with either a hot or cold high-pressure water cleaner at least 24 h before use and immediately after the pig has been removed. Optionally, cleaning agents can be used, but local environmental regulations need to be checked to determine which agents are permitted. Prior to pigs entering the metabolic cages, the water supply should be checked to confirm that it is working, so that pigs have free access to water whilst in the metabolic cage (Fig. 5.1a and c). If water metres have been installed, they should be checked for proper functioning, so that water intake by the pigs can be recorded.

How long an animal stays in the metabolic cage will depend on your experimental design and aims. As some general guidelines, under EU Regulation: Directive 2010/63/EU (for scientific use), there is no maximum time that pigs can be held in metabolic cages. Instead, the system is classified by severity, as follows:

- Short-term (< 24 h) is classified as severity degree 1 (non-harmful/mild).
- Moderate durations (up to 5 days) are classified as severity degree 2 (moderate).
- Long-term (> 5 days) are classified as severity degree 3 (severe).

The Swiss Animal Protection Ordinance also includes a minimum recovery period for stays in metabolic cages, depending on how long the animals are kept in a metabolic cage and the level of stress caused by the experiment (Federal Food

Safety and Veterinary Office, 2017). For a stay in metabolic cages of up to 8 h, the minimum recovery period is at least 16 h. For stays from 8 to 24 h, the minimum recovery period is 6 days. For stays from 4 to 7 days, the minimum recovery period must be the length of the detention period plus 14 days.

There are also no EU regulations regarding space allowance in a metabolic cage. Rather, space allowance is determined by the size of the pigs and the study design. Some studies require the movement of pigs to be restricted to standing and sitting. This requires that the sides, front and back of the metabolic cage are adjusted inwards so that the pig cannot move forward, backwards or turn around. This is important when urine needs to be collected while minimizing the potential for any faecal contamination, for example, in nitrogen balance studies, where the nitrogen excreted in the urine is quantified. Another example would be studies that use metabolic cage-mounted infusion pumps. During these studies, pigs receive continuous intravenous infusions, and it is critical that the animals are unable to pull the catheter out (Rasch et al. 2021). However, such a restriction is only recommended for the infusion period. Once the infusion period is complete, the sides of the metabolic cage should be adjusted, allowing the pig more movement. In other studies, pigs do not need to be restrained in the metabolic cage. An example would be studies where data on faecal and urine output needs to be collected whilst blood (ATOL_0005631) samples are collected from indwelling catheters that exit on the back of the pig and are fixed to the pig by bandaging to prevent them from being pulled out (Cavalleri et al. 2025). In general, when contamination of urine with faeces is a major concern, the use of males (castrated or uncastrated) is recommended over females, since it is easier to minimise the contamination of urine with faeces in males.

Important
In the end, the type and size of the metabolic cage, as well as the identity of subjects and severity grade and duration of restriction, will be determined by the researchers based on the goals of their study. However, we recommend that researchers design their studies to minimise the time that pigs spend in metabolic cages while ensuring robust data collection. Any metabolic cage study beyond 5 days constitutes severe restriction by EU law and requires rigorous ethical justification and oversight.

5.3.2.2 Animal Adaptation to the Metabolic Cages

Habituation to the metabolic cage is vitally important to the health and welfare (ATOL_0000765) of the animals used in any study, as animals that are well adapted have reduced levels of stress, and can be easily handled during the experimental period. Here we provide some examples of habituation techniques, based on studies conducted by our research groups. However, we would like to emphasise that these are only examples. The adaptation protocols used for any metabolic cage study will vary depending on the study's specific requirements. For example, in a study with catheterised pigs, weighing between 20 and 40 kg, the pigs were first housed in individual stalls with transparent walls, so that the pigs could see each other. The individual stalls contained enrichment devices (EOL_0001921, balls, chew toys, ropes, etc.), and twice a day, staff entered the stalls to check catheters and play with the pigs. The pigs were then habituated to the metabolic cages three times over a 1-week period. On habituation days, the subjects were moved to metabolic cages at 7:00 and kept in the cages until 16:00. Pigs were then fed their morning meals at 7:30 in the cages.

As another example, in a balance study, urine was collected from the same animal at two different stages over 5 consecutive days: once during the growing period (around 40 kg) and once during the finishing period (around 80 kg). Before placing the animals in metabolic cages, they were kept in groups for at least two weeks. During this time, the animals habituated to their companions and to the feed. Any pigs that were reluctant to interact with the experimenter were not selected for the cages. On the day that they were placed in the metabolic cages, several pigs from the same enclosure entered their cages in the morning. Once in the cage, a small food reward was given, followed by the morning feed ration. The first day in the cages was used for additional habituation, and no samples were taken. During their stay in the cages, the pigs received one meal in the morning and one meal at the end of the day. An experimenter checked on the pigs two to three times during the day to ensure that they were eating and drinking properly.

5.3.2.3 Preparation of Urine Collection Containers

- The collection containers should be clean (not necessarily sterile) and leak-proof and preferably have a lid or mesh cover to avoid contamination.
- The materials and surfaces in contact with the urine should neither react with nor adsorb any analytes of interest (Delanghe and Speeckaert 2014; Kiyokawa et al. 2011). In addition, they should be free of interfering substances and particles. For instance, albumin binds to the surface of some plastics, which may lead to poor recovery at low concentrations (Hara and Shiba 2003). Plastic (e.g. polyethylene or polypropylene) or glass containers may be used for urea or nitrogen analyses. Additional requirements, such as amber-coloured containers to analyse light-sensitive analytes (such as porphyrins and urobilinogen, for instance), may be needed.
- If required, carefully label the collection containers with an indelible marker. The labels should resist water, urine and/or acid.
- In some studies, such as balance studies, it is important to know the exact volume and/or weight of urine collected and to use a preservative solution. For this, it is advisable to prepare the collection containers 1 or 2 days before sampling begins.
  - Prepare a spreadsheet containing the weight of all empty containers. Decide whether or not to include the weight of the cap in this calculation.
  Then, prepare the preservative solution (if required). For example, 3 M sulphuric acid can be used to acidify urine in order to determine its nitrogen content. Add 25 g of this solution to the containers and note the exact weight on the sheet (accuracy: 0.001 g). As concentrated sulfuric acid is highly corrosive, it

should always be diluted by adding the acid to water rather than the other way around. Wear gloves, safety goggles and a lab coat, and use acid-resistant containers.

5.3.2.4 Preparation of Urine Sub-Sample Collection Tubes/ Containers

- Carefully label the sub-sample collection tubes/containers with an indelible marker. The labels should resist water, urine and/or acid.
- If the study requires a urine sub-sample that is representative of the entire collection period (e.g. 2% *w/w* of the total daily urine excreted), create a spreadsheet to calculate the necessary urine weight to be aliquoted.

5.3.3   Description of the Procedure

5.3.3.1 Daily Sample Collection

1. Mount the urine collection tray under the metabolic cage.
2. Set up the urine collection system, by setting up the filter system and then the collection container(s) below the collection tray (Fig. 5.2a–c).
3. According to the metabolites that need to be analysed, it may be necessary to:
4. Place a refrigerated system below the collection container (e.g. with an icebox that might be changed frequently according to the room temperature— Sect. 5.3.2).
5. Place a container containing a preservative solution (Sect. 5.3.2).
6. If urine is being collected throughout the day, check the volume collected once or twice a day. If the container is two-thirds full, wear lab gloves and replace it with a spare one that is clean and dry, to prevent dilution of the urine content.
7. Record the volume and/or the weight of urine collected on a spreadsheet if this is required.

5.3.3.2 Collection over Several Days

In particular, for balance studies, urine must be collected over several consecutive days (e.g. 5 days). For the first collection day, follow Sect. 5.3.3.1, steps 1–5. On the second day:

1. Remove the containers from the collection system.
2. Clean the metabolic cages with water; this helps to limit contamination with faeces. Pigs remain in the cages during the cleaning.
3. Replace the full urine collection containers with new, clean, dry containers.
4. Weigh all the containers (non-acidified and acidified, if applicable) to determine the amount of urine produced in 24 h. Determine the amount to be aliquoted (e.g. 2% w/w of the total urine produced) and record it on the spreadsheet.

5. Gently shake the container to ensure that the urine is mixed evenly. Weigh the aliquots in 0.5–1 litre containers (Sects. 5.3.1.4 and 5.3.2.4), and freeze them immediately at −20 °C.
6. Once the aliquots have been collected, wash the urine collection containers with hot water and dry them with paper towels or a cloth so that they can be used the next day.
7. On the subsequent collection days, follow the same steps as on the second day of collection. After weighing the urine, place the aliquots in the 0.5–1 litre containers (one per pig) that were frozen the previous day and freeze them, thus creating a pooled sample for each pig.
8. Check the volume of urine produced once or twice a day. If the containers are two-thirds full, remove them, close them with a cap and leave them in the larger container (see the grey container (3) in Fig. 5.2a). Then, replace them with clean, new containers.

5.3.3.3 Sub-sampling and Storage

1. Collected urine samples should be mixed by holding the container and moving it around to ensure that they are homogeneous.
2. The urine should then be pipetted into new clean tubes (Sects. 5.3.1.4 and 5.3.2.4) and centrifuged according to the experimental protocol required. For the measurement of ^{13}C enrichment, lactulose and mannitol, urine samples have been centrifuged at $1000 \times g$ for 10 min at 4 °C. For the measurement of cortisol (ATOL_0005350), creatinine, amino acids, amino-metabolites, nitrogenous waste products and markers of protein catabolism, urine samples have been centrifuged at $3000 \times g$ for 10 min at room temperature. Centrifuging helps to remove particulate matter that could interfere with sample measurement.
3. Supernatants should then be sub-sampled to tubes with an appropriate volume for planned downstream analyses (Sect. 5.3.1.4). Downstream analyses may include the measurement of: (1) ^{13}C enrichment using isotope-ratio mass spectrometry (Cavalleri et al. 2025); (2) lactulose, mannitol, amino acids and/or amino-metabolites using HPLC; (3) cortisol or other hormones using enzyme-linked immunosorbent assays; or (4) creatinine, markers of protein catabolism and/or nitrogenous waste products using colorimetric and enzymatic assays.
4. If tubes are to be frozen, fill them up to two-thirds of the total tube volume to prevent them from bursting due to volume expansion during freezing. We recommend storing at −20 °C for up to 3 months and at −80 °C when storing for more than 3 months.
5. To obtain a representative sample of the total collection period when collecting urine over several consecutive days, weigh the amount of urine produced per day (Sect. 5.3.3.2) and collect a fixed percentage (e.g. 2% *w/w*). Since the amount of urine produced varies from day to day, this allows for standardisation. Record these amounts (total amount produced and amount collected) on a spreadsheet, then freeze the sample at −20 °C. Then, fill this same tube with the subsequent days' aliquots, i.e. use a single tube for all days as described in Sect. 5.3.3.2. Just

before analysis, thaw the aliquots, collect a portion in a new tube (the volume depends on the analysis) and centrifuge it as described above.

5.4 Manual Urine Collection by Spontaneous Urination from Unrestrained Pigs

5.4.1 Materials and Equipment

5.4.1.1 Housing

Pigs can be housed either individually or in groups. The size of the pens depends on each country's legislation. Examples of housing for pigs in different categories, both individually and in groups, are shown in Fig. 5.3. Pigs must have ad libitum access to fresh drinking water for the entire duration of the experiment. They prefer drinking from a trough (EOL_0001618), rather than a nipple drinker. Be sure to change the water regularly especially in the summer. Also, free range sows must have access to a mud hole, when temperatures reach above 15 °C. Otherwise, they might bathe in the water trough and contaminate the drinking water.

Fig. 5.3 Examples of pig housing systems where the manual urine collection by spontaneous urination method can be used: (**a**) free-range system for lactating sows (Credit: Maria Eskildsen, Aarhus); (**b**) individual system for lactating sows (Credit: Marion Girard, Agroscope); (**c**) group system for fattening pigs (Credit: Marion Girard, Agroscope)

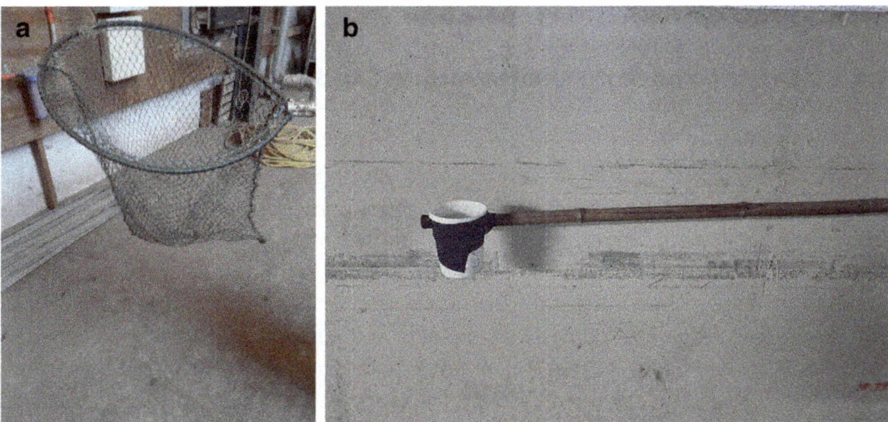

Fig. 5.4 Examples of home-made urine collection systems: with a fishing net (**a**) in which a collection bucket can be placed, or a stick with the collection cup attached with adhesive tape (**b**) (Credit: Maria Eskildsen, Aarhus and Marion Girard, Agroscope)

5.4.1.2 Materials and Equipment for Urine Sample Collection

- Urine should be collected in a bucket or cup with a minimum capacity of 500 mL. If the animal is some distance away, homemade devices such as a net to hold the bucket, or a stick with a collection cup attached (Fig. 5.4), can make collection easier.
- Clean collection containers (preferably sterile).
- Long-handled ladle or scoop.
- Disposable gloves.
- Protective clothing and boots.
- Labels and waterproof markers.
- Cooler with ice packs for sample storage.
- Disinfectant for cleaning equipment.

5.4.1.3 Materials and Equipment for Weighing and Urine Sub-Sample Preparation

- Refer to Sect. 5.3.1.4 for details of the materials required for sample processing, including weighing (if needed) and sub-sampling.
- Eppendorf tubes or 50–100 mL tubes/containers for the sub-samples.
- Urine preservatives may need to be added for analysis of specific metabolites.
- For example, to stabilise pH and prevent volatilisation when nitrogen is analysed, we recommend adding 0.5 mL of sulfuric acid at 1 M to 50 mL urine sub-sample tubes. This results in a pH < 2, which stabilises some urinary metabolites, such as nitrogen, when this is necessary for specific analyses.
- Pipettes and tips (volumes will depend on how much is being sub-sampled), to prepare the sub-samples.

5.4.2 Prerequisites and Preparation

5.4.2.1 Description of Different Housing Conditions

Pigs may be housed in free-range housing systems under certified organic conditions in accordance with EU regulations (EU 2018/848). During gestation, several sows may be housed together in common huts. During lactation, each sow should be housed individually with her litter in outdoor farrowing huts, which provide shelter, thermal protection and the opportunity for undisturbed maternal behaviour. These huts are distributed across spacious paddocks with continuous access to pasture, enabling the expression of natural behaviours such as foraging, rooting and exploration. Each free-range unit consists of one sow and her piglets, to prevent inter-sow aggression and to facilitate natural nursing and bonding. Outdoor paddocks are maintained with sufficient vegetation cover to support voluntary intake of grass, and the layout of the facilities allows for reliable assessment of feed intake (ATOL_0000772), energy expenditure through locomotive activity and thermoregulatory demands (ATOL_0000857). The water trough should be full and fresh drinking water should be available at all times.

When working with lactating free-range sows, EU law (EU 2018/848) requires that huts are arranged in such a way that the animals can see each other. Huts should be closed in advance before dawn, as the sows will urinate first thing in the morning. Close the huts quietly, as too much noise will wake them up and possibly cause them to urinate within the hut. When collecting urine from pregnant sows in common huts, it is important to have enough stock personnel (one person per sow) or to adjust the number of sows that are let out of the hut at the same time to the number of people available. The sows will likely urinate at the same time, and one person will not be able to collect from more than one sow. It is very time-consuming to collect urine from outdoor pigs at times of the day other than in the morning. Urine can also be collected from pigs in more conventional housing conditions. As with free-range animals, it is advisable to be present in the morning to collect the samples, since animals often urinate shortly after waking up.

5.4.2.2 Animals

Urine can be collected from males and females of any age. If it is not possible to collect in the morning, it is advisable to monitor the pig's behaviour and anticipate urination, which will often happen after feeding or drinking. The procedure is non-invasive and free of pain for the animals. However, some pigs might be stressed by being followed by a person for 5–10 min. To reduce the stress level of the animals and minimise the risk of attack (especially in lactating sows), it is recommended that pigs become habituated to the experimenters. To achieve this, 1 to 2 weeks before collection begins, experimenters should spend at least 10–15 min a day in the pens with the animals, observing them. When the animals allow them to approach, the experimenters should try to stroke, pat and interact positively with them. Observing the animals for longer periods can also help to identify where and when they urinate. The experimenter may also enter the pen with the container that will be used for collection and attempt to take samples, which will then be discarded. If an animal shies away from human contact, the collection container can be mounted on a 2-metre-long stick with adhesive tape or be placed inside a fishing net (Fig. 5.4). If this system is to be used, it must be introduced into the pens during the habituation period and tested before the collection phase begins. If pigs are reluctant to urinate, it can help to lead them gently to the defaecation (ATOL_0000492) area where they usually urinate.

5.4.2.3 Preparation of Urine Collection Containers and Sub-Sample Tubes

The bucket or collection container (cup with or without fishnet/stick method) should be clean (not necessarily sterile) and leak-proof.

The bucket or collection container and the sub-sample tubes should be labelled with an indelible marker, which can resist being rubbed off by movement (such as that experienced when carrying the bucket behind the sow before she urinates), water, urine and/or acid.

If the volume or weight of the collected urine needs to be measured, the weight of each container must be recorded.

If the metabolite you analyse requires a chemical preservative after collection (e.g. for nitrogen analysis), put it in the sub-sample tubes. For example, by adding 0.5 mL of sulfuric acid at 1 M to 50 mL urine sub-sample tubes as mentioned in Sect. 5.4.1.3. As concentrated sulfuric acid is highly corrosive, it should always be diluted by adding the acid to water rather than the other way around. Wear gloves, safety goggles and a lab coat, and use acid-resistant containers.

5.4.3   Description of the Procedure

5.4.3.1 Sample Collection

The concentration of nutrients and urine volume can vary considerably depending on weather, dry matter in the feed (grass/roughage), amount of water drunk, etc. As previously mentioned, it is also recommended to collect the sample when the pig wakes up, as it is more likely that pigs will urinate soon after waking up.

5.4.3.2 Procedure

> **Important**
> Do not collect urine from pigs alone, as there are risks of injury if animals become aggressive (ATOL_0000914). However, thorough habituation can help to minimise this risk.

1. Take the bucket or the collection container. If needed, set up the urine collection system (e.g. mount the collection stick and then the collection container below).

2. Have enough staff available in case the animals all urinate at the same time. It is also a good idea to have one person responsible for processing samples while the others collect urine.

   a) In a free-range system, one person should open the hut while others are assigned to each sow. After opening the hut, each person should follow a pig at a 2–3-metre distance. The pigs will most often urinate within the first five minutes after leaving the hut in the morning (Fig. 5.5a).

   b) Indoors, pigs kept in groups often urinate in the same area of the pen. In this case, it is sufficient to follow the pigs at a distance of 1–2 metres, especially when they approach this area.

Fig. 5.5 Manual urine collection by spontaneous urination method in free-range sows (**a**) and fishing net used to hold bucket with container for urine collection in group housing (**b**, **c**) (Credit: Maria Eskildsen, Aarhus)

3. When a sample has been collected and the collection bottle needs to be changed between animals, wear lab gloves (nitrile or latex) and remove the collection bottle. Replace with a new collection bottle that is clean and dry to prevent dilution of the contents.

5.4.3.3 Sub-sampling and Storage

- Mix the urine in the collection bottle by shaking the bottle, and then use a pipette to transfer around 1 or 2% of the total amount of collected urine into labelled tubes of 2–100 mL.
- We recommend centrifuging samples at 3000 × g for 10 min to remove particulate matter. Then, aliquot the supernatant into sterile tubes.
- Several tubes can be used if several analyses are required.
- If tubes are to be frozen, fill them only up to two-thirds of the total tube volume, to prevent them from bursting due to volume expansion during freezing. We recommend storage at −20 °C for up to 3 months and −80 °C for more than 3 months.

5.5 Compliance with the 3R Principles

The metabolic cage protocol outlined in this SOP requires that the pigs be restricted for an extended period with minimal movement, which is likely to negatively impact on their welfare. However, metabolic cages are currently one of the only techniques that allow for repeated and continuous non-invasive urine collection from one animal, whereas this is not currently practical in the spontaneous urination system. Metabolic cages thus reduce the number of animals required for studies, where the exact volume of urine produced per animal is required, since alternatives under open housing systems require using a larger number of animals to obtain accurate estimates of daily urine volumes. In addition, multiple tests can be performed in the metabolic cages using the same animal, further reducing the number of animals required and recycling those currently in the study.

The manual urine collection by the spontaneous urination method is currently the only technique that allows completely non-invasive urine collection of pigs in a more natural environment. This method is useful when one or two daily samples are sufficient to capture changes in analytes of interest. However, the method is less useful for continuous collection of urine, i.e. throughout the day, as this requires enough staff to be on site at all times, which is often impractical. In addition, the constant staff presence may disrupt the natural behaviour of the animals. Your experimental protocol needs to clearly justify the nature of the sample collection required and thus the method to be used (i.e. metabolic crate vs. manual urine collection by spontaneous urination).

Table 5.1 Advantages and disadvantages of using metabolic cages or open environments for urine sampling

Method	Advantages	Disadvantages
Metabolic cages	Easy separation of urine and faeces Can perform repeated and continuous sampling of urine Non-invasive Enables total daily collection	Acclimation needed Risk of faeces contamination when using female animals Welfare concerns due to lack of space and social isolation
Manual urine collection by spontaneous urination	Easy separation of urine and faeces Non-invasive No restraint needed Allows sampling in home environments and in social groups	Time- and personnel-intensive Difficult to continuously sample animals Greater risk of physical injury to personnel

5.6 Conclusions

This SOP outlines two minimally or non-invasive procedures for collecting urine. While this SOP is primarily based on examples from metabolic studies, it is also applicable to other fields where the analysis of analytes in urine is relevant. Metabolic cages are currently one of the only techniques that allow continuous, sequential sample collection, whereas manual urine collection by spontaneous urination allows for sporadic collection under more natural conditions. The advantages and disadvantages of the procedures outlined in this SOP are summarised in Table 5.1.

Acknowledgments The authors of this SOP would like to thank Dr. Susan A. McCoard (AgResearch, Palmerston North, New Zealand) and Dr. Sarah Pain (Massey University, Palmerston North, New Zealand) for their expert help in preparing this document.

References

Berghaus D, Haese E, Weishaar R et al (2023) Nitrogen and lysine utilization efficiencies, protein turnover, and blood urea concentrations in crossbred grower pigs at marginal dietary lysine concentration. J Anim Sci 101:skad335. https://doi.org/10.1093/jas/skad335

Cavalleri M, Sciascia M, Görs S et al (2025) Measurement of oro-cecal transit time in LPS-treated pigs fed high and low fiber diets using the lactose-13C-ureide test in breath and saliva samples. J Agric Food Chem 73(17):10304–10315. https://doi.org/10.1021/acs.jafc.5c00534

Czech A, Samolińska W, Tomaszewska E et al (2022) Effect of microbial phytase on ileal digestibility of minerals, plasma and urine metabolites, and bone mineral concentrations in growing–finishing pigs. Animals 12(10):1294. https://doi.org/10.3390/ani12101294

Delanghe J, Speeckaert M (2014) Preanalytical requirements of urinalysis. Biochem Med 24(1):89–104. https://doi.org/10.11613/BM.2014.011

Dhondt L, Croubels S, De Paepe P et al (2020) Conventional pig as ani-mal model for human renal drug excretion processes: unravelling the porcine renal function by use of a cocktail of exogenous markers. Front Pharmacol 11:883. https://doi.org/10.3389/fphar.2020.00883

Federal Food Safety and Veterinary Office (2017) Technical information 2.06: Keeping laboratory animals in metabolism cages and boxes. https://www.blv.admin.ch/dam/blv/en/dokumente/tiere/publikationen-und-forschung/tierversuche/richtlinien-stoffwechselkaefig.pdf.download.pdf/fachinformation-2-06-halten-von-versuchtieren-stoffwechselkaefige-stoffwechselboxen_EN.pdf. Accessed 9 Oct 2025

Hara F, Shiba K (2003) Nonspecific binding of urinary albumin on preservation tube. Jpn J Clin Chem 32(Suppl 1):28–29

Jia Y, Hwang SY, House JD et al (2010) Long-term high intake of whole proteins results in renal damage in pigs. J Nutr 140(9):1646–1652. https://doi.org/10.3945/jn.110.123034

Kim YJ, Kim TH, Song MH et al (2020) Effects of different levels of crude protein and protease on nitrogen utilization, nutrient digestibility, and growth performance in growing pigs. J Anim Sci Technol 62(5):659–667. https://doi.org/10.5187/jast.2020.62.5.659

Kiyokawa I, Sogawa K, Ise K et al (2011) Adsorption of urinary proteins on the conventionally used urine collection tubes: possible effects on urinary proteome analysis and prevention of the adsorption by polymer coating. Int J Proteomics:502845. https://doi.org/10.1155/2011/502845

Nixon E, Mays TP, Routh PA et al (2020) Plasma, urine and tissue concentrations of Flunixin and meloxicam in pigs. BMC Vet Res 16:1–10. https://doi.org/10.1186/s12917-020-02556-4

Rasch I, Görs S, Tuchscherer A et al (2021) Substitution of dietary sulfur amino acids by DL-2-hydroxy-4-Methylthiobutyric acid increases Remethylation and decreases Transsulfuration in weaned piglets. J Nutr 149(3):432–440. https://doi.org/10.1093/jn/nxy296

Sachse D, Solevåg AL, Berg JP et al (2016) The role of plasma and urine metabolomics in identifying new biomarkers in severe newborn asphyxia: a study of asphyxiated new-born pigs following cardiopulmonary resuscitation. PLoS One 11(8):e0161123. https://doi.org/10.1371/journal.pone.0161123

Svoboda M, Nemeckova M, Medkova D et al (2024) Non-invasive methods for analysing pig welfare biomarkers. Vet Med (Praha) 69(5):137–155. https://doi.org/10.17221/17/2024-VETMED

Tkaczyk A, Jedziniak P, Zielonka Ł et al (2021) Biomarkers of deoxynivalenol, citrinin, ochratoxin a and zearalenone in pigs after exposure to naturally contaminated feed close to guidance values. Toxins 13(11):750. https://doi.org/10.3390/toxins13110750

Faeces Collection and Procedure for Determination of Apparent Total Tract and Ileal Digestibility Using Indigestible Markers

6

Marion Girard ⓘ, Sonja de Vries, Emma Ivarsson, Virág Ács, Tamme Zandstra, Nikkie van der Wielen, Carolien De Cuyper, Helle Nygaard Laerke, and Sam Millet

Abstract

To accurately formulate diets that meet pigs' nutritional requirements, the digestibility of energy and nutrients of feed ingredients should be estimated. Apparent total tract digestibility (ATTD) and apparent ileal digestibility (AID) are two important indicators for evaluating the nutritional content of feed ingredients and diets. The present guideline proposes a standard operating procedure (SOP) for measuring ATTD and AID during the slaughter procedure using an indigestible marker. The SOP outlines how to prepare the necessary materials, diets, and animals for the experiment, as well as how to collect, process, store, and analyse faecal and ileal samples from individual pigs, including how to calculate the ATTD and AID of a diet. The procedure is customised for pigs weighing ≥ 15 kg housed in groups or individual pens. Unlike the total collection method, which requires total digesta collection, the index marker method allows partial digesta or faeces sampling. This approach reduces the need for individual housing in metabolic cages, as pigs can be housed in groups. The procedure does not include details on the method of euthanising the animal for AID determination. Each

Supplementary Information: The online version contains supplementary material available at https://doi.org/10.1007/978-3-032-24899-2_6.

M. Girard (✉)
Swine Research Group, Agroscope, Posieux, Switzerland
e-mail: marion.girard@agroscope.admin.ch

S. de Vries · T. Zandstra
Animal Nutrition Group, Wageningen University and Research (WUR), Wageningen, The Netherlands

E. Ivarsson
Department of Applied Animal Science and Welfare, Swedish University of Agricultural Sciences (SLU), Uppsala, Sweden

researcher must comply with the ethical rules in force in the country in which the study is conducted.

Keywords

Apparent ileal digestibility · Marker · Pig · Total tract digestibility

6.1 Introduction

To meet the maintenance and production needs of pigs while minimising feed costs and reducing the environmental impact of pig production, an appropriate supply of dietary energy and nutrients is crucial (Zhang and Adeola 2017). To accurately formulate diets that meet these nutrient requirements, the digestibility of energy (ATOL_0001828) and the nutrients (ATOL_0001231) of feed ingredients should be estimated (Kong and Adeola 2014). The determination of the nutrient and energy digestibility of feedstuffs is a simple method for characterising the availability of nutrients in the ingredients included in pig diets (Sauer and Ozimek 1986).

Apparent total tract digestibility (ATTD) (ATOL_0001232) and apparent ileal digestibility (AID) are two important indicators for evaluating nutritional or energy value in feed ingredients and diets. Accurate estimation of dietary energy (EOL_0002011) is an essential prerequisite in pig nutrition because energy represents the main feed cost and is used to predict feed intake (ATOL_0000772) and, subsequently, the requirements for other nutrients (Noblet and Van Milgen 2004; Kil et al. 2013).

Although the net energy (EOL_0001935) reflects the actual available energy for pigs, the digestible energy (EOL_0001933) and the metabolisable energy (EOL_0001934) are easier to estimate because the methodology for evaluating these indices uses faeces or faeces and urine rather than calorimetry. Therefore, ATTD is still widely used to determine the energy values of feed ingredients or diets. However, the site of digestion and absorption of amino acids in pigs is the

V. Ács
Department of Farm Animal Nutrition, Hungarian University of Agricultural and Life Sciences (MATE), Kaposvár, Hungary

N. van der Wielen
Animal Nutrition Group, Wageningen University and Research (WUR), Wageningen, The Netherlands

Nutritional biology group of the Division of Human Nutrition and Health, Wageningen University and Research (WUR), Wageningen, The Netherlands

C. De Cuyper · S. Millet
Animal sciences Unit, Flanders Research Institute for Agriculture, Fisheries and Food (ILVO), Merelbeke-Melle, Belgium

H. N. Laerke
Department of Animal and Veterinary Sciences, Aarhus University, Tjele, Denmark

small intestine. Amino acids that reach the large intestine are degraded or converted into other amino acids or metabolites by fermentation and, therefore, do not provide essential amino acids to the pig. Thus, the evaluation of amino acid digestibility (ATOL_0001247) is based on the collection of digesta samples from the end of the ileum.

AID can be determined by estimating the amount of amino acids ingested and the amount present in the ileum. However, digesta samples from the ileum contain not only undigested feed protein but also the undigested protein of endogenous origins, such as undigested digestive enzymes, proteins from desquamated epithelial cells, mucus, or bacterial proteins. Ileal endogenous losses may be separated into basal losses, which are not influenced by feed ingredient composition, and specific losses, which are induced by feed ingredient characteristics, such as dietary protein content (EOL_0000124), levels and types of fibre, and anti-nutritional factors. The correction of the AID values for both basal and specific endogenous losses leads to true ileal digestibility (ATOL_0001258). However, reliable procedures to routinely measure specific endogenous losses are not yet available. Given that the basal endogenous losses are related to the total dry matter intake (ATOL_0005395), it is possible to correct the AID values for the excretion of undigested basal endogenous protein, thereby yielding standardised ileal digestibility values. The determination of endogenous losses remains a matter of debate. Table values can be used to assess these losses, which reduces the number of animals required and the experimental costs. However, evidence shows that these losses are study-specific and may vary according to the experimental conditions, and thus, table values may not reflect the losses associated with a specific context and individual animal (Blok et al. 2022). Hence, the use of nitrogen-free feed is recommended in all studies to determine endogenous losses (Blok et al. 2022; Stein et al. 2007).

Two methods are used for in vivo digestibility studies: the total collection method and the index marker (IM) method. The classic total collection is laborious and time-consuming because it requires a quantitative collection and record of feed intake and total tract faeces excretion to determine the difference between the components in ingested feed and excreted faeces. Failing to fully collect all faeces can cause potential errors, which may result in the overestimation of digestibility. To facilitate total collection, metabolic cages (EOL_0002023) are primarily used for this method. After 5–21 days of adaptation to cages and feeds (Lyu et al. 2018; Zhao et al. 2018), a total collection of faeces is carried out for 4–6 days (Zhang and Adeola 2017). Total collection is considered the gold standard method for determining digestibility, but it requires a labour-intensive recording of feed intake and faeces output (Kong and Adeola 2014). The IM is a less labour-intensive method (Kavanagh et al. 2001) and allows partial sampling of faeces and feed. This method does not require metabolic cages (EOL_0002023) if the goal is solely to measure digestibility. However, IM requires precise and accurate chemical analysis of indigestible markers. Various indigestible markers have been proposed, such as titanium dioxide (TiO_2), chromium oxide (Cr_2O_3), ferric oxide (Fe_2O_3), and acid-insoluble ash (AIA or silicon dioxide [SiO_2]). A proper indigestible marker should be: (Brestenský et al. 2017; Zhang and Adeola 2017):

(a) Totally indigestible and non-absorbable
(b) Non-toxic to the host and its digestive tract
(c) Able to track the component of interest through the digestive tract
(d) Easily analysed accurately and precisely
(e) If possible, a compound/substance that is naturally present in the feed

6.2   Goal and Scope of the Procedure

The present guideline proposes an SOP to measure ATTD and AID by the slaughter procedure using an indigestible marker. It describes how to prepare the materials, the diets, and the animals for the experiment (Sect. 6.3 and 6.4), how to collect, process, store, and analyse faecal and ileal samples (Sects. 6.5.1 and 6.5.2) from individual pigs, and how to calculate ATTD and AID of a diet (Sect. 6.6).

This procedure is adapted to pigs weighing $\geq$15 kg housed in groups or individual pens. For AID determination, this procedure does not describe the method for euthanising the animal. Each experimenter must comply with the ethical rules in force in the country in which the trial is being carried out.

6.3   Materials and Equipment

6.3.1   Faeces Collection on Live Pigs

- Laboratory/examination gloves (powder-free).
- Nitrile or vinyl: preferred to latex as they are less allergenic.
- Water-resistant felt-tip pen or freezer-proof labels.
- Spoons or plastic shovel: to collect a certain quantity of faeces.
- Plastic or aluminium foil trays/plates, small buckets, or plastic bags to collect faeces.
- Cotton swab: to stimulate defaecation (ATOL_0000492) in small pigs (if needed).
- Styrofoam container with ice: to transport the sub-samples (if needed).
- Pig scale (accuracy 100 g): to weigh the pigs (if needed for the experiment).

6.3.2   Ileal Digesta Collection after Slaughter

- Laboratory gloves/examination gloves (powder-free).
- Nitrile or vinyl: preferred to latex as they are less allergenic.
- Water-resistant felt-tip pen or freezer-proof labels.
- Plastic container with lid: to collect ileal content (at least 200 mL capacity).
- Pig scale (accuracy 100 g): to weigh the pigs (if needed for the experiment).
- Styrofoam container with ice: to transport the sub-samples (if needed).
- Dissection material for ileum collection (tie-wraps, medical lancet, medical pliers, scissors, for instance).

6.3.3   Sample Processing and Storage

- Inert recipient (~2–5 L) for pooling the faeces (option 1 in Sect. 6.5.1.4).
- Mixer, if faecal (e.g. paddle or hand-held kitchen mixer) or digesta (e.g. stirring plate) samples are pooled (option 1 in Sect. 6.5.1.4).
- Plates or metal trays for freezing the samples. They should also be resistant to drying.
- Certified electric scale (accuracy 100 mg): to weigh the quantity of faeces and ileal content (if dry matter (DM) needs to be determined).
- Freezer (-20 °C/-80 °C): before freeze-drying the digesta, it is important to ensure that it is thoroughly frozen.
- Freeze-drier or oven.
- Grinder: to grind freeze-dried and dried faeces, ileal, and feed samples.

6.4   Prerequisites and Preparation of the Experiment

6.4.1   Preparation of the Feed and the Indigestible Marker

6.4.1.1 Choice of the Indigestible Marker

As mentioned earlier, a marker must meet certain criteria, such as being completely indigestible and non-absorbable, non-toxic, able to pass through the digestive tract at a constant rate with the digesta, easy to analyse, and preferably being naturally present in the feed. Many studies have compared different markers and identified unique advantages and disadvantages. The choice of marker depends on various factors, such as local legislation, the purpose of the study, or the analytical methods available in the laboratory (de Vries and Gerrits 2018). Two commonly used markers (SiO_2 and TiO_2) are described in Table 6.1. The limiting factor that determines the inclusion level of the marker is its quantification in the feed rather than its quantification in the faeces and digesta, where it is highly concentrated. To be reliably quantified, the marker must be incorporated into the feed in sufficient quantity to reduce measurement inaccuracies when levels are close to the detection and quantification limits of the analytical method. In addition, the marker should minimise confounding effects of trace quantities of the substance naturally present in feed ingredients (e.g. in the case of AIA), without being too concentrated in the digesta to avoid interference with digestive processes.

6.4.1.2 Feed Preparation and Analysis

- The feed including the marker can be offered in dry (mashed, flour, pelleted) or liquid form.
- In the case of leftovers in the feeders, it is important to check that there is no segregation between the marker and the diet.
- In trials in which the ATTD of phosphorus is determined, the basal diet should consist of raw materials without endogenous phytase activity.

Table 6.1 Chemical composition of two commonly used markers and their inclusion levels in a diet

Applied marker	Inclusion level range in diet (as-fed basis)	Remarks
Silicon dioxide (SiO_2; AIA)	0.50–1.5%	The high inclusion level of AIA may impair digestive processes (passage rate) for highly digestible diets, where the concentration of AIA gets very high in ileal content. A larger amount of sample is required for chemical analysis. Intrinsic and added AIAs may have uneven particle sizes, which can lead to improper mixing and differential passage rates. To ensure consistent particle size and even distribution, a source of fine-particle AIA or an inert material, such as Celite™ is recommended (Goddard and McLean 2001).
Titanium dioxide (TiO_2)	0.10–0.40%	Too high inclusion level ($\geq 0.50\%$) may affect palatability and faecal consistency. Environmental and health issues: Under current EFSA regulations, the use of TiO2 (and other external markers that are not registered as feed additives, including Cr_2O_3) as an additive is prohibited (EU, 2021/2090), but its use as a tracer in scientific experiments can be authorised by the local authority based on EU directives (87/153/EEC and 83/228/EEC). Animals monitored with TiO_2 must not re-enter the food chain. TiO_2 may interact with phosphorus (Kuboki et al. 2012).

- Depending on the nutrients of interest in the study, the chemical composition of the diet, such as DM, crude ash, crude protein, crude fibre, ether extract, and minerals, may be analysed before starting the trial.

6.4.1.3 Feed Allowance and Feeding Frequency

- The feed can be provided *ad libitum* or restricted. Feeding the pigs restricted but close to *ad libitum* to avoid feed residues is recommended, especially if feed is not pelleted, and to avoid diverging growth rates between treatment groups. Moreover, as feeding level may affect digestibility, different diets must be compared at the same feeding level. However, for group-housed pigs, *ad libitum* feeding may be more convenient to avoid competition.
- It is suggested that the feeding level should be approximately 2.8 times the maintenance energy requirement or approximately 4% of live body weight (ATOL_0000351) (Zhang and Adeola 2017; CVB 2005), given that nutrient digestibility could be significantly influenced by low feed intake (Moter and Stein 2004).
- In cases of restricted feeding, the feed allowance should be provided twice daily in two equal meals, with a minimum interval of 8 h between meals, or even in three equal meals for young pigs.
- If pigs are feed-restricted, the daily feed allowance may be calculated using the following equation (Eq. 6.1; Everts 2015):

$$Daily\ feed\ allowance(kg) = \frac{LW^{0.60} \, x \, 750 \, x \, FL}{NE_{diet}} \tag{6.1}$$

where:

- LW is the live weight measured at the beginning of a specific research period (e.g. adaptation or collection, see Sect. 6.4.3 and 6.5.1.2).
- $LW^{0.60}$ is the live weight to the power of 0.60.
- 750 is the energy for maintenance in kJ NE/day per kg live weight to a power of 0.60 (Everts 2015).
- FL is the feeding level; this coefficient should be approximately 2.3–2.8 times the maintenance level for net energy (NE).
- NE_{diet} (in kJ/kg) is the estimated net energy content of the diet.

6.4.2 Animals and Housing

6.4.2.1 Housing Conditions

- Animals can be either housed in groups or individually, must have *ad libitum* access to water for the entire duration of the experiment, and must be identifiable (ear tag or number).
- No straw or other bedding should be allowed in the pens in the second phase of the adaptation period or in the collection period (please refer to Sect. 6.4.3 for the definition of the phases of the adaptation period).
- The pigs may have access to inedible and non-destructible enrichment materials (EOL_0001921, e.g. metal chains with hard plastic attributes or large balls) during the whole experiment to encourage exploratory behaviour. However, some enrichment materials, such as jute products containing cellulose, should be avoided.

6.4.2.2 Animals

- Faeces or ileal digesta can be collected from all sexes (castrated boars, boars, or females) and from all physiological stages (piglets, fattening pigs, sows). If samples are taken from young animals, the challenge is to obtain enough material for analysis; otherwise, pools of samples from several animals should be made.
- If this SOP is used to determine the nutrient values of a test ingredient for feeding tables, please refer to the specific feeding system guidelines for the recommended weight of the animals at the time of measurement.
- It is important to define the experimental unit before the start of the trial. Depending on the configuration of the facility, animals kept in a group in the same pen may either receive the same dietary treatment (e.g. pen with one feeder) or different dietary treatments (e.g. pen with several individual electronic feeders, each dispensing one feed or pen with one individual electronic feeder dispensing multiple feeds). Thus, the experimental unit may be the pen or the animal, respectively.
- The number of replicates depends on the experimental unit (animal or pen), the variable studied (ATTD or AID and the nutrient of interest), and the differences between treatments that the experiment aims to detect (power calculation). A

greater number of replicates is required for diets with high fibre content (Pedersen et al. 2010). To determine the number of replicates, please refer to the guidelines of De Cuyper et al. (2025a, 2025b). The CVB (2005) protocol recommends that the standard deviation of the ATTD of organic matter (OM) should be less than 1.5%. Typically, four to six replicates are used for standard diets.

- For animal welfare (ATOL_0000765) reasons, it is important to socialise pigs. The purpose of socialisation is to accustom pigs to human contact, ensuring they remain calm during tasks such as pen cleaning and sample collection. Interactions between pigs and humans involve acoustic, visual, tactile, and chemical sensory channels. The animals must become familiar with people's smells and voices, and they must allow themselves to be petted. Socialisation can begin one or two days after the pigs have become accustomed to their new environment, that is, on days 2 or 3 of the first adaptation phase (see Sect. 6.4.3 for details of the first phase).

Important

This is an example of a socialisation protocol performed in different steps at Wageningen University:

1. Whenever you enter the room, talk to the pigs or whistle to help them become accustomed to your presence and the sound of your voice.
2. Enter the pen and talk to each pig. Do not make any sudden movements. Just sit and talk to the pig to let it get used to you being in the pen in a non-threatening manner. Spend 5–10 min with each pig.
3. Enter the pen, and after 2 min of sitting and talking, extend your hand for the pig to sniff and explore. Extend your arm to pet the pig. Let the pig get used to this, and then, try keeping your hand on its back. The pig will probably run around the pen. Do not remove your hand until the pig stops moving.
4. Get closer to the pig, and keep your hand on its back for longer. Start giving the pig scratches on its back, behind its ears, and on its belly. You can also bring a container to collect faeces to help them get used to it.
5. Optional: If the pigs are still small, you can try holding them (wear earplugs). This is often the most difficult step, as they will often scream. Do not release the pig unless it stops squealing. Being quiet is the key to being released.
6. Optional: Keep trying to hold the pigs every day until they submit to being held without squealing. From this point onwards, they should always allow you to hold them and rub their belly and back.

Typically, each day, the next step is attempted. Some pigs may take longer to socialise. Others may take less time. If they are quicker, some steps can be combined and performed in one day. During the adaptation period, you can reward the pigs with treats, such as fruit (apples or dates) or chocolate raisins, after completing a step. As pigs like playing, you can also freeze ice blocks in buckets containing pelleted feed and give the pigs these ice blocks during the socialisation period. Then during the measurement/sampling process, you can continue to provide the ice blocks, but without any feed in them.

6.4.3 Adaptation Period to the Diet

A period of adaptation must precede each collection period to adapt the animals to the new diet and environment. This period is also important for achieving homogeneous mixing of the marker in the digestive tract. The duration of the period can vary according to the diet composition, the feed allowance (restricted or *ad libitum*), and whether pigs need to adapt to a new environment. This adaptation period can be divided into two parts:

- *First phase:* This first phase is recommended if pigs need to be housed in a new environment, are housed individually, need to be fed restrictedly during the collection period, or if a feed change needs to take place gradually. This first phase of adaptation lasts at least 5 days.
- *Second phase:* This second phase adapts the intestinal physiology to the new feed and allows the marker to reach a steady excretion rate from the digestive tract. This phase begins when the animals have received only the diet to be studied. For example, at least 5 days were required before the marker excretion stabilised after feeding a corn-soybean meal-based diet containing chromic oxide (Cr_2O_3) in Jang et al.'s (2014) study. However, the duration of this second phase may vary according to the type of marker and the composition of the studied diet, particularly the digestible non-starch polysaccharide content. The adaptation period must be extended for diets containing a high content of low-digestible compounds, such as dietary fibres, to avoid a discrepancy in the intestinal passage rate (Clawson et al. 1955). According to the CVB (2005) protocol, 5 days are necessary for diets containing less than 50 g/kg DM of digestible non-starch polysaccharides, but this period may extend to more than 10 days for diets containing more than 150 g/kg DM of digestible non-starch polysaccharides or even to 14 days when determining the digestibility of phosphorus. It is therefore advisable to have a second phase adaptation period of at least 5 to 7 days for low-fibre diets and of at least 14 days for high-fibre diets or the determination of phosphorus digestibility.

6.4.4 Preparation of the Material for Faeces Collection

- The collection plates must be identifiable and weighed before the experiment if DM needs to be determined (Supplementary Material 6.1 for an example of a recording sheet).
- If pigs are housed in groups, a second person may be needed to keep other pigs from disturbing the pig or the person taking the sample.

6.5   Description of the Specific Procedures

6.5.1   Procedure for Determining Apparent Total Tract Digestibility by Faeces Collection on Live Pigs

6.5.1.1 Weighing Dates
- Animals should always be weighed at the same time of day.
- The following main weighing dates may be considered:

  - At the start of the adaptation period, especially if animals are fed restrictedly, to adjust the amount of feed.
  - 1 to 3 days before the end of the adaptation period (to avoid stress (ATOL_0002301) during the collection period), especially if animals are fed restrictedly, to adjust the amount of feed
  - At the end of the collection period.

6.5.1.2 Duration of the Collection Period
- It is important to collect faecal samples over several days before pooling them together to obtain more robust results. It is also important to ensure that each collection day is equally represented. A sample collection period that preferably lasts 3–5 days is recommended.
- Pooling grab samples over two consecutive days has been shown to result in greater accuracy and lower variations than a single grab sample (Jang et al. 2014). According to Moughan et al. (1990), a sampling period of 5 days allows the precision of IM for determining DM digestibility to approach that of the total collection method. More precise determination of nutrient digestibility with five collection days was also reported by Agudelo et al. (2010) and Brestenský et al. (2017), whereas Jang et al. (2014) did not observe any improvement by pooling faeces collected across 2–5 days. A sample collection period that preferably lasts 3–5 days is recommended.

6.5.1.3 Faeces Collection.
- Faeces must be collected at least once a day, preferably at the same time every day, using a plate, a small bucket, or a bag.
- Wear laboratory gloves during sampling. If the experimental unit is:
  - The pig: change gloves between pigs
  - The pen: change gloves between pens
- Samples must be collected immediately after defaecation on the floor or directly from the rectum. If the pen is the experimental unit, collect faeces from at least 2 animals per pen.
  - Defaecation is induced by gently stimulating the final 3–6 cm of the rectum (depending on the animal's size) with slow, circular movements of the index finger. Alternatively, a cotton swab can be used to collect faeces rectally on pigs weighing less than 20 kg. It is important to insert the finger or cotton

swab gently and not abruptly. The sample can be taken without restraining the animal, but if it moves too much, it can be held briefly by a second person. Sampling takes a few seconds to a few minutes (less than 2 min).

- If faeces are collected from the floor, only collect the part that has not touched the floor.
- If AIA is used as a marker, it is important to collect the sample rectally to avoid contamination with AIA from the environment.
- After collection, faecal samples must be frozen immediately or stored at 4 °C for a maximum of 4 hours before being frozen.

- The quantity depends on the duration of the collection period; the more collection days, the smaller the quantity to be collected per day. The DM of faeces typically varies between 190 and 280 g/kg; thus, it is advisable to obtain approximately 500 g of faeces per experimental unit at the end of the collection period.
- If there is no defaecation at the time of sampling, returning a little later or omitting this sample is recommended. However, it is important to have at least three collection days per experimental unit.
- Make sure that all containers holding faeces are labelled correctly before storage.

6.5.1.4 Faeces Homogenisation and Storage

Following collection, there are two ways to proceed for homogenisation and storage, as described below:

- Option 1: Pooling at the end of the collection period (Fig. 6.1).

  1. Immediately after collection, transport the faecal samples on ice in the Styrofoam container, freeze them, and store them at −20 °C until the end of the collection period.

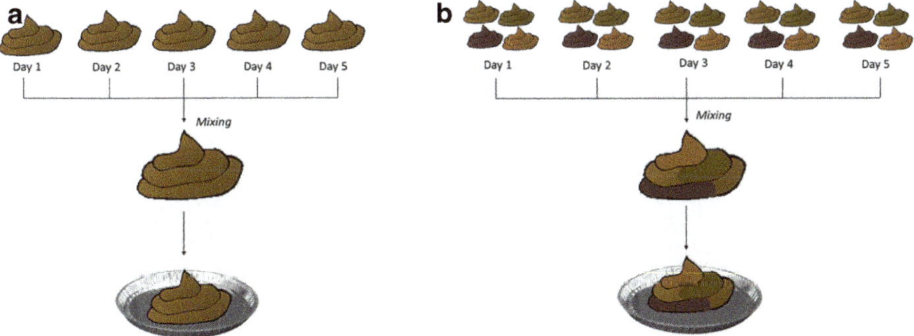

Fig. 6.1 Pooling at the end of the experimental collection period. Defrosted faeces samples from either the same pig (individual level; a) or from several pigs of the same pen (pen level; b) are pooled before freeze-drying

2. Once all faeces samples have been collected and frozen, defrost the samples in closed buckets or bags by leaving them at 4 °C for a maximum of two days to minimise microbial activity.
3. After defrosting, pool the samples from each experimental unit (animal or pen) within each experimental collection period by adding the same quantity of faeces per day from the same animal/pen in a suitable container. If the pen is the experimental unit, the faeces of at least two animals per pen should be pooled.
4. Homogenise the pooled sample using a mixer to ensure a representative sample. Clean the mixer between samples.
5. A subsample of approximately 250–500 g of homogenised pooled faeces is then taken and placed on a tray or plate suitable for freeze-drying or oven-drying. The exact quantity of the subsample is recorded before freeze-drying or oven-drying if DM needs to be determined. If this subsample is to be freeze-dried, freeze it at −20 °C (it could also be pre-frozen at −70 °C). Samples intended to be dried in an oven should be processed immediately. A second subsample may be collected and frozen as a backup.

- Option 2: Daily pooling during the experimental collection period (Fig. 6.2).

1. On the first day of the experimental period, labelled plates resistant to freeze-drying or oven-drying (in aluminium or resistant plastic for instance) are tared (one plate per experimental unit), and the weight of the empty plates is recorded. Shortly after collection, a certain amount of faeces is weighed. A total of 100 g per experimental unit per day is optimal. If the experimental unit is the pig, 100 g of faecal sample is weighed. If the experimental unit is the pen, a lower amount of faeces is required, and the faeces of at least two animals per pen are collected to reach 100 g. If a large number of faecal samples are to be collected, one person can collect the faeces, while another weighs them and places them on the plates. As collecting all the samples may take some time, the plates can be kept cool in a freezer or in a Styrofoam container filled with ice. All the plates are then frozen and stored at −20 °C.

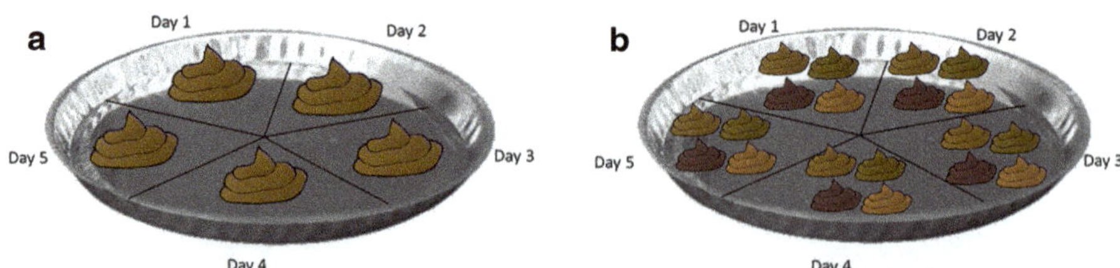

Fig. 6.2 Daily pooling throughout the experimental collection period of freshly collected faecal samples from either the same pig (individual level; a) or from several pigs of the same pen (pen level; b) before freeze-drying

2. On the subsequent day of the experimental period, the plates containing the frozen faeces from the previous day are taken out of the freezer and kept in a Styrofoam container filled with ice to avoid defrosting. Fresh faeces are collected. The plate containing the frozen faeces from the same experimental unit as the previous day is tared. The same quantity of freshly collected faeces as the previous day is added to this tared plate, that is, 100 g per experimental unit, for instance. The exact weight of the fresh faeces for the new day is recorded. If a large number of faecal samples are to be collected, one person can collect the faeces, while another weighs them and places them on the plates. As collecting all the samples may take some time, the plates can be kept cool in a freezer or a Styrofoam container filled with ice. All the plates are then returned to the freezer at $-20\ °\mathrm{C}$.

3. Step 2 is repeated for the remaining collection days.

4. At the end of the experimental collection period, if DM needs to be determined, the weights of the faeces collected over the days are added to obtain the total weight of the faeces in the plate.

6.5.1.5 Feed Collection and Storage

- Representative samples of feed must be collected during the experimental collection period to analyse the chemical composition and the indigestible marker content. Ideally, a feed sample should be collected on each collection day and then pooled at the end of the collection period for chemical analyses.
- Individual feed intake may be recorded according to the purpose of the experiment, but this record is not mandatory in the IM method.

6.5.1.6 Laboratory Analysis

- The pooled faecal samples, that is, either pooled at the end of the experimental collection period (Sect. 6.5.1.4 Option 1 and Fig. 6.1) or pooled daily during the experimental collection period (Sect. 6.5.1.4 Option 2 and Fig. 6.2), should be freeze-dried or oven-dried until reaching a constant weight and then weighed if the DM content needs to be determined. Afterwards, the freeze-dried or oven-dried samples are ground in order to pass through a 0.5 to 1 mm grid according to the subsequent gross chemical analysis (Kavanagh et al. 2001; Battelli et al. 2020; Prawirodigdo et al. 2021).
- Feed samples must be oven-dried (48 h at 60 $°\mathrm{C}$) or freeze-dried if the moisture content exceeds 15% (*w/w*), then ground to pass through a 0.5 to 1 mm grid.
- Accredited and/or validated methods must be used to determine the DM and indigestible marker content in both lyophilised/oven-dried feed and lyophilised/oven-dried faecal samples.
- Depending on the purpose of the experiment, the nutrients of interest (e.g. gross energy, phosphorus, and calcium) are also determined in the faeces and the feed using accredited and/or validated methods.

6.5.2   Procedure for Determining Apparent Ileal Digestibility by Ileal Digesta Collection after Slaughter

6.5.2.1 Recommendations before Slaughter

- The slaughter technique has been shown to be an alternative to simple T-cannulation and a suitable technique for measuring nitrogen (ATOL_0001829) and amino acid digestibility in pigs receiving meat and bone meal diets (Donkoh et al. 1994). However, this technique, which provides an instantaneous 'picture', may result in more variations than cannulation, which samples over a longer period (Donkoh et al. 1994; Pedersen et al. 2010). Furthermore, it is more challenging to collect the requisite quantity for analysis from a single animal using the slaughtering technique.
- The slaughter day can take place the week following the end of the adaptation period.
- Animals are weighed on the day of slaughter.
- Unlike normal slaughter conditions, animals should not be fasted to ensure that sufficient ileal content is collected. Therefore, 48 to 72 h prior to dissection, the meal frequency can be increased to four to six meals evenly distributed at regular intervals, and on the day of dissection, pigs are fed once every hour and receive at least six meals (one meal is 1/12th of the daily allowance) to approach steady-state digesta passage as described by van Erp et al. (2020). Alternatively, on the day of slaughter, the morning meal may be divided into several portions, and euthanasia must then take place a minimum of 6 h after the first portion is consumed, and a maximum of 1 h after the last portion.
- Containers for collecting ileal content should be resistant to freeze-drying, and weighed if DM determination is needed.

6.5.2.2 Ileal Sample Collection and Storage

- The collection of ileal content occurs post-mortem.
  - Only one pig should be euthanised at a time, and the contents of the ileum should be collected immediately afterwards.
- Make sure that all containers holding ileal content are labelled correctly.
- A qualified person should perform the anaesthesia or stunning (e.g. electro-stunning, CO_2, captive bolt pistol) before the euthanasia of the pigs.
  - If the sole purpose of the experiment is to collect content from the ileum, a stunning method may be employed. However, if additional samples need to be taken from different parts of the intestine, or if intestinal morphology or kinetics are to be studied, then anaesthesia must be used.
- After euthanasia, the whole digestive tract is carefully collected immediately.
  - To prevent the movement of the digesta upstream, we recommend tying one string at the beginning of the duodenum (in close proximity to the stomach) and one at the termination of the ileum, in the immediate vicinity of the caecum, prior to the unfolding of the entire small intestine.

- The beginning of the caecum is identified, and then the mesentery is gradually and carefully removed until the small intestine is unfolded.
  - The ileum is located in the last 4–5% of the small intestine. As the length of the small intestine varies according to the age of the pig (from 7 m at weaning to over 16 m for an adult pig; Laerke and Hedemann 2012), as well as between pigs of the same age, we recommend that the entire length of the small intestine be unfolded and measured, and the contents should be collected from the final 5% (anterior to the ileocecal valve), omitting the last 5–10 cm from the ileocecal valve.
- The content (digesta) is collected from the final 35 or 80 centimetres (excluding the last 5–10 cm close to the ileocecal valve) in the case of piglets or adult pigs, respectively, within 10 min after euthanasia in identified containers resistant to freeze-drying.
  - To optimise the freeze-drying process, the contents of the ileum in the freeze-drying-resistant container are spread to a maximum height of 2–3 cm. If the height is excessive, a dry 'crust' may form and prevent the sample from freeze-drying completely. The sample could also be pre-frozen at −70 °C before the freeze-drying process.
  - If DM determination is needed, the containers should have been previously tared, and the digesta weight of the experimental unit is recorded with a 0.1 g precision scale.
  - Since chemical analyses are performed on freeze-dried samples and the DM of ileal samples varies between 70 and 140 g/kg, collecting at least 250 g of ileal content per experimental unit is recommended. When the experimental unit is the pen, the ileal contents of several animals from the same pen (at least two animals) are mixed in the same container. Ideally, the same quantity should be combined per animal within the same experimental unit. When the experimental unit is the pig, the ileal content of a single animal may not be sufficient. In such instances, it is necessary to take the entire ileal content of a second animal that has received the same treatment and combine it with the initial sample to create a pool.
- Immediately after sampling the experimental unit, the container with the digesta is kept cool in a Styrofoam container filled with ice. Once all samples have been collected and processed, they are transported on ice before being frozen and stored at −20 °C or − 70 °C.

6.5.2.3 Feed Collection and Storage

Samples of feed must ideally be collected during the last days prior to slaughter, particularly if the feed is not pelleted, and pooled to analyse the chemical composition and the indigestible marker content.

6.5.2.4 Laboratory Analysis

Once frozen, the ileal content samples must be freeze-dried (generally for 72 h) until they reach a constant weight and then weighed if the DM content needs to be determined.

Please refer to Sect. 6.5.1.6 for the grinding of freeze-dried ileal samples, the processing of the feed samples, and the determination of the different DMs and indigestible marker content in the feed and the ileal samples.

Depending on the purpose of the experiment, the nutrients of interest (e.g. specific minerals and specific amino acids) are also determined in the faeces and the feed using accredited and/or validated methods.

6.6   Calculation Method for Digestibility

The ATTD of nutrient N is calculated from the concentration of the nutrient in the feed (N_{feed}) and in the faeces (N_{faeces}) and the concentration of the indigestible marker in the feed (M_{feed}) and the faeces (M_{faeces}) with the formula (Eq. 6.2; Adeola 2001):

$$\mathrm{ATTD_N}\,(\%) = 100 - \left[100 \times \frac{\mathrm{M_{feed}}}{\mathrm{M_{faeces}}} \times \frac{\mathrm{N_{faeces}}}{\mathrm{N_{feed}}} \right] \tag{6.2}$$

M_{feed}, M_{faeces}, N_{feed}, and N_{faeces} are expressed in g/kg DM.

The AID of nutrient N is calculated from the concentration of the nutrient in the feed (N_{feed}) and the ileum (N_{ileum}) and the concentration of the indigestible marker in the feed (M_{feed}) and the ileum (M_{ileum}) (Eq. 6.3):

$$\mathrm{AID_N}\,(\%) = 100 - \left[100 \times \frac{\mathrm{M_{feed}}}{\mathrm{M_{ileum}}} \times \frac{\mathrm{N_{ileum}}}{\mathrm{N_{feed}}} \right] \tag{6.3}$$

M_{feed}, M_{ileum}, N_{feed} and N_{ileum} are expressed in g/kg DM.

Due to variations in the analysed concentration of the marker that cannot be explained by dosing errors, it is advisable to check whether the analysed values of the test diets are within the analytical margin of error. If this is the case, it is advisable to take an average value for all diets with the same anticipated marker concentration as those fed to the animals. In the case of AIA, as intrinsic AIA content may differ among test ingredients and thus diets, taking an average AIA value per diet is recommended.

6.7   Compliance with the 3R Principles

The determination of apparent total tract and ileal digestibility using indigestible markers complies with the 3R principles. However, compliance with European, national, and local ethical rules and regulations should always be ensured. The use of the present SOP contributes to reducing stress and discomfort in pigs. In both procedures described, the animals may be housed in groups, unlike other procedures that require metabolic cages or ileal cannulated pigs. The pigs may also have inedible and non-destructible enrichment material to enable explorative behaviour during the trial.

The determination of ATTD using an indigestible marker is minimally invasive because faeces can be easily collected from the rectum of the animals in their pens or when defecating. Socialising and habituating pigs to human contact during the adaptation period are key steps in reducing animal stress and refining faecal collection during the experimental period. Furthermore, the ATTD method can be used in cross-over studies, reducing the number of animals required in an experiment, as faeces can be collected from the same pig at different collection periods. However, for some nutrients, such as amino acids, ileal digestibility must be determined, and ileal digesta must be collected. Post-mortem collection can be an alternative to cannulation, and the collection of ileum samples may be combined with the collection of other organs and tissues (liver, large intestine, etc.).

6.8 Conclusions

Assessing the nutrient digestibility of feed ingredients and diets is a prerequisite for optimising diet formulation and thereby pig growth and health (ATOL_0000928). It is also an effective tool for tackling environmental issues aimed at lowering nitrogen and phosphorous excretions. Unlike the total collection method, which requires total digesta collection, the IM method described in this SOP allows partial digesta sampling. This method reduces the need for individual housing in metabolic cages during the experimental period, as pigs can be housed in group pens, which is more in line with their natural behaviour. Collecting ileal samples post-mortem can also replace the use of cannulated pigs, but different diets cannot be tested on the same animal.

Several compounds can be used as indigestible markers. However, indigestible markers containing heavy metals, such as chromium oxide or titanium dioxide, can cause environmental and suspected health issues. As an alternative, AIA can be used but must be included at a higher level due to the sensitivity of the analytical method used to determine insoluble ash content as well as to ensure proper mixing of the different particles between intrinsic and added AIAs.

Acknowledgements The authors would like to thank Professor Ruurd Zijlstra (University of Alberta, CA) and Dr. Karl Schedle (Boku University, AU) for reviewing and improving earlier versions of this SOP.

References

Adeola O (2001) Digestion and balance techniques in pigs In: Lewis AJ. Southern LL (eds) Swine nutrition, 2nd edn. CRC Press, Washington, pp 903–916

Agudelo J, Lindemann M, Cromwell G (2010) A comparison of two methods to assess nutrient digestibility in pigs. Livest Sci 133:74–77. https://doi.org/10.1016/j.livsci.2010.06.029

Battelli M, Rapetti L, Rota Graziosi A et al (2020) Use of undigested NDF for estimation of diet digestibility in growing pigs. Animals 10(11):2007. https://doi.org/10.3390/ani10112007

Blok M, Spek W, Bikker P (2022) P104 basal endogenous amino acid losses in the digestive tract are determined by experimental conditions, a meta-analysis. Anim Sci Proc 13(2):188–189

Brestenský M, Nitrayová S, Heger J et al (2017) Chromic oxide and acid-insoluble ash as markers in digestibility studies with growing pigs and sows. J Anim Physiol Anim Nutr 101(1):46–52. https://doi.org/10.1111/jpn.12503

Clawson A, Reid J, Sheffy B et al (1955) Use of chromium oxide in digestion studies with swine. J Ani Sci 14:700–709. https://doi.org/10.1093/ansci/14.3.700

CVB (2005) CVB (Centraal Veevoederbureau) Protocol voor een fecale verteringsproef met groeiende intacte vleesvarkens. https://wwwcvbdiervoedingnl/pagina/10622/protocollenaspx

De Cuyper C, Ampe B, Westin R et al (2025a) Opinion paper: improving experimental design in animal research – seven key principles. Animal 19(12):101695. https://doi.org/10.1016/j.animal.2025.101695

De Cuyper C, Westin R, Millet S et al (2025b) PIGWEB guidelines for experimental design. Zenodo. https://doi.org/10.5281/zenodo.15371453

de Vries S, Gerrits WJJ (2018) The use of tracers or markers in digestion studies. In: Moughan PJ, Hendriks WH (eds) Feed evaluation science. Wageningen Academic Publishers, pp 275–294

Donkoh A, Moughan PJ, Smith WC (1994) Comparison of the slaughter method and simple T-piece cannulation of the terminal ileum for determining ileal amino acid digestibility in meat and bone meal for the growing pig. Anim Feed Sci Technol 49(1–2):43–56. https://doi.org/10.1016/0377-8401(94)90080-9

Everts H (2015) Energy requirement for maintenance in growing pigs (CVB-documentation report; no nr 57). WUR Livestock Research Dept Animal Nutrition. https://edepotwurnl/375603

Goddard JS, McLean E (2001) Acid-insoluble ash as an inert reference material for digestibility studies in tilapia, Oreochromis aureus. Aquaculture 194(1–2):93–98. https://doi.org/10.1016/S0044-8486(00)00499-3

Jang YD, Lindemann MD, Agudelo-Trujillo JH et al (2014) Comparison of direct and indirect estimates of apparent total tract digestibility in swine with effort to reduce variation by pooling of multiple day fecal samples. J Ani Sci 92(10):4566–4576. https://doi.org/10.2527/jas.2013-6570

Kavanagh S, Lynch P, O'Mara F et al (2001) A comparison of total collection and marker technique for the measurement of apparent digestibility of diets for growing pigs. Anim Feed Sci Technol 89:49–58. https://doi.org/10.1016/S0377-8401(00)00237-6

Kil D, Kim B, Stein HH (2013) Feed energy evaluation for growing pigs. Asian Australas J Anim Sci 26:1205. https://doi.org/10.5713/ajas.2013.r.02

Kong C, Adeola O (2014) Evaluation of amino acid and energy utilization in feedstuff for swine and poultry diets. Asian Australas J Anim Sci 27(7):917–925. https://doi.org/10.5713/ajas.2014.r.02

Kuboki Y, Furusawa T, Sato M et al (2012) Interaction between titanium and phosphoproteins revealed by chromatography column packed with titanium beads. Biomed Mater Eng 22(5):283–288. https://doi.org/10.3233/BME-2012-0718

Lærke HN, Hedemann MS (2012) The digestive system of the pig (chapter 5) In: Bach Knudsen KE, Kjeldsen NJ, Poulsen HD, Jensen BB (eds) Nutritional physiology of pigs. Online publication, Videncenter for Svineproduktion: Foulum Denmark

Lyu ZQ, Huang CF, Li YK et al (2018) Adaptation duration for net energy determination of high fiber diets in growing pigs. Anim Feed Sci Technol 241:15–26. https://doi.org/10.1016/j.anifeedsci.2018.04.008

Moughan PJ, Smith WC, Schrama J et al (1990) Chromic oxide and acid-insoluble ash as faecal markers in digestibility studies with young growing pigs. N Z J Agric Res 34:85–88. https://doi.org/10.1080/00288233.1991.10417796

Moter V, Stein HH (2004) Effect of feed intake on endogenous losses and amino acid and energy digestibility by growing pigs. J Ani Sci 82(12):3518–3525. https://doi.org/10.2527/2004.82123518x

Noblet J, Van Milgen J (2004) Energy value of pig feeds: effect of pig body weight and energy evaluation system. J Ani Sci 82:229–328

Pedersen C, Vinyeta E, Gerritsen R (2010) Comparison of the slaughtering and the T-cannulation methods on ileal digestibility of nutrients in pigs. In: Crovetto GM (ed) energy and protein metabolism and nutrition. Wageningen Academic Publishers, EAAP Scientific Series 127(1):411–412

Prawirodigdo S, Gannon NJ, Leury BJ et al (2021) Acid-insoluble ash is a better indigestible marker than chromic oxide to measure apparent total tract digestibility in pigs. Anim Nutr 7(1):64–71. https://doi.org/10.1016/j.aninu.2020.07.003

Sauer WC, Ozimek L (1986) Digestibility of amino acids in swine: results and their practical applications. A review. Livest Prod Sci 15(4):367–388. https://doi.org/10.1016/j.aninu.2020.07.003

Stein HH, Sève B, Fuller MF et al (2007) Invited review: amino acid bioavailability and digestibility in pig feed ingredients: terminology and application. J Ani Sci 85(1):172–180. https://doi.org/10.2527/jas.2005-742

van Erp RJJ, de Vries S, van Kempen TATG et al (2020) Feed intake patterns nor growth rates of pigs are affected by dietary resistant starch despite marked differences in digestion. Animal 14(7):1402–1412. https://doi.org/10.1017/S1751731119002945

Zhang F, Adeola O (2017) Techniques for evaluating digestibility of energy amino acids phosphorus and calcium in feed ingredients for pigs. Anim Nutr 3(4):344–352. https://doi.org/10.1016/j.aninu.2017.06.008

Zhao J, Zhang S, Xie F, Li D, Huang C (2018) Effects of inclusion level and adaptation period on nutrient digestibility and digestible energy of wheat bran in growing-finishing pigs. Asian Australas J Anim Sci 31:116–122. https://doi.org/10.5713/ajas.17.0277

Sample Handling Procedure for Determination of Ileal Digestibility by Near-Infrared Spectroscopy

7

Samantha Joan Noel, Knud Erik Bach-Knudsen, David Renaudeau, Josep Ramos, Rosil Lizardo, Sonja de Vries, Sam Millet, George Bazar, Mátyás Krisztián Lukács, Haruna Gado Yakubu, and Jan Værum Nørgaard

Abstract

Near-infrared spectroscopy (NIR) is a rapid, cost-effective, and environmentally friendly method for evaluating the nutrient content of feedstuffs. NIR analysis can be performed on various sample types, including feed samples (both raw materials and mixed diets) and samples collected from animals, such as faeces or

S. J. Noel (✉) · K. E. Bach-Knudsen · J. V. Nørgaard
Department of Animal and Veterinary Science, Aarhus University (AU), Foulum, Denmark
e-mail: samantha.noel@anivet.au.dk

D. Renaudeau
PEGASE, INRAE, Insitut Agro, Saint-Gilles, France

J. Ramos · R. Lizardo
Animal Nutrition Program, Institut de Recerca i Tecnologia Agroalimentàries (IRTA), Constantí, Spain

S. de Vries
Animal Nutrition Group, Wageningen University & Research (WUR), Wageningen, The Netherlands

S. Millet
Animal sciences Unit, Flanders Research Institute for Agriculture, Fisheries and Food (ILVO), Merelbeke-Melle, Belgium

G. Bazar
ADEXGO Kft. Correltech Laboratory, Budapest, Hungary

M. K. Lukács
Department of Food Measurement and Process Control, Institute of Food Science and Technology, Hungarian University of Agriculture and Life Sciences (MATE), Budapest, Hungary

H. G. Yakubu
Department of Animal Science, School of Agriculture, University of Cape Coast, Cape Coast, Ghana

© The Author(s) 2026

L. R. Moscovice et al. (eds.), *Standard Operating Procedures for Better Pig Research (SOPig)*, https://doi.org/10.1007/978-3-032-24899-2_7

digesta content. It does not require reagents, thus producing no waste, and preserves the sample while estimating multiple parameters simultaneously. NIR relies on product-specific calibrations developed from samples measured with chemical or physical reference methods. Predictions of feedstuff digestibility made by NIR have shown accuracy comparable to or exceeding other proxy measurements, reducing the need for animal experimentation, supporting the 3Rs. The goal is to predict ileal apparent amino acid digestibility values accurately, rapidly, and cost-effectively. This SOP describes the procedure to obtain reliable NIR scans and predictions from feedstuffs and faeces. In conclusion, NIR analysis is fast, non-destructive, and cost-effective without involving animal testing. However, it is indirect and depends on strong, validated prediction models, with reliable results only when samples match the model's variation range.

Keywords

NIRS · Feedstuff evaluation · Digestibility prediction · Apparent amino acid digestibility · Rapid analysis

7.1 Introduction

Ensuring the feed meets the animal's needs is pivotal to efficient animal production and to minimise environmental and climate impacts (Millet et al. 2018; Wang and Zijlstra 2018). Historically, this is accomplished in practise by using table values to calculate diet formulations, or with time-consuming and expensive animal trials to determine digestibility (Patience 2018). Table values for nutrient digestibility (ATOL_0001231) are based on average values calculated from digestibility experiments with animals and do not account for individual variation in diet or diet ingredients or animal responses (Knudsen et al. 2023). Alternatively, the nutrient composition and value of feed can be measured by wet chemical methods and in vitro techniques, which are time-consuming and expensive, limiting routine use (Boisen and Fernández 1997; Noblet and Perez 1993).

Near-infrared spectroscopy (NIR) offers a rapid, cost-effective, and environmentally friendly method for evaluating feedstuffs (Baeten et al. 2016). It can be used to estimate the nutritional composition of animal feeds as well as nutritional value, e.g. the digestibility of the feed and specific feed components, thus offering a quick, inexpensive and reliable proxy for standard measurements and animal trials (Bastianelli 2013; Chen et al. 2011; Fontaine et al. 2002; Noel et al. 2021, 2022). NIR on faeces can give predictions of individual animal nutrient digestibility that can be used to select pigs with improved nutrient utilisation (Bastianelli et al. 2015; Schiborra et al. 2015). NIR is based on the interaction of light with organic matter, and it does not require reagents and therefore does not produce waste. The sample is not destroyed, and several parameters can be estimated simultaneously, with each scan taking roughly 1 minute.

NIR is an indirect method that relies on product-specific calibrations developed from samples measured with chemical or physical reference methods. Predictions of feedstuff digestibility made by NIR have been shown to be as or more accurate than other proxy measurements (Noel et al. 2021, 2022). Thereby, NIR is a way to reduce the need for animal experimentation as it can be used for estimating chemical composition and biological characteristics such as the digestibility of feedstuffs and complete feeds.

During scanning, light from the near infrared wavelength range (780 to 2500 nm) is irradiated on the sample and is partly absorbed and partly reflected. The reflected light is captured by sensitive detectors, converted into a signal, and recorded. Within a sample, absorption of light in the NIR region causes molecules to vibrate and rotate, and if the vibration of the molecules is of the same frequency as the exposure wavelength, an amplified signal is reflected. Thus, infrared spectra contain quantitative information about functional groups containing the covalent bonds of C-H, O-H, and N-H. However, this is complicated by the overlapping of spectral bands originating from overtones and other molecules. Therefore, NIR spectroscopy is an indirect or secondary technique, which requires chemometric modelling to relate the complex spectra to reference values. Samples used to generate the calibration must be representative of the possible variation within the product in question, for which the parameters to be predicted were determined with chemical reference methods. Once developed, the calibration model can be used to predict values from new samples without any reference measurements.

7.2   Goal and Scope of the Procedure

This SOP describes the procedure for how to prepare and store samples for NIR analysis and the scanning of samples with the aim of predicting ileal apparent amino acid digestibility (ATOL_0001247) values from either pig feed or faeces. As some parts of the procedure are specific to the type of NIR scanner employed, this protocol will only detail general sample handling and scanning procedures. Also, it does not include detailed information on how to collect feed or faecal samples or the number of samples needed for a certain type of experiment. For more information on faecal sampling, see Chap. 6.

7.3   Materials and Equipment

7.3.1   Sample Preparation

- Freeze dryer (e.g. Lyotech 5 (Coolvacuum, Granollers, Spain)
- Grinder (e.g. Ultra Centrifugal Mill ZM 200 (Retsch, Haan, Germany) or MultiDrive basic (IKA, Staufen, Germany) equipped with a 0.5 mm or 1 mm grid)
- Airtight sample containers for storing samples

7.3.2 NIR Scanning

- NIR scanning machine (e.g. FOSS DS2500, scanning range 400–2500 nm)
- NIR machine software (e.g. ISIscan Nova software or IQX)
- Calibration models by chemometric software (e.g. Foss Calibrator, Matlab, or R)

7.4 Prerequisites and Preparation

NIR analysis can be performed on a variety of sample types, including feed samples (both raw materials and mixed diets) and samples collected from animals, e.g. faeces samples or digesta content. A suitable calibration model must first be loaded into the NIR machine software. These models may be purchased as products from the NIR supplier or developed in-house from libraries of samples with measured reference values. The type of sample to be measured must fall into the range of the intended prediction model, i.e. be the same type of sample that the model was developed with. Each calibration model will be developed from a known sample group with measured reference values. For generating or updating calibrations, it is preferable to perform the NIR scan on the reference samples within 1 month of reference chemical analysis; however, successful calibrations have been made on samples that have been bio-banked for several years if they are stored correctly (dry at $-20\ ^{\circ}C$ in airtight containers). Previously developed calibration models can be updated by adding new reference samples with measured reference values. Make sure the instrument maintenance and calibration are up to date before scanning your experimental samples.

7.5 Description of the Scanning Procedure

7.5.1 Sample Preparation and Storage

- NIR scan can be performed on either animal feed or faeces to estimate the ileal digestibility of the amino acids in the feed.
- After the sample (feed or faeces) has been collected, a homogenisation of the total sample is performed before a representative sample for drying and grinding is taken.
- The amount of fresh faeces to collect to obtain the minimal amount of freeze-dried material depends on the water content of the faecal material but >100 g is recommended. The aim is to have enough material to cover the bottom of the 7 cm wide NIR sample cup (Fig. 7.1), which requires, depending on the sample type, ~ > 25 g or ~ 3 heaped tablespoons or ~ 60 mL of dried material.
- Faeces samples can be from a single spot/grab collection from a live animal or at slaughter or from a pooled collection per animal taken over 2–3 days (see Chap. 6 for examples of collection and pooling). Once all days are collected, the pooled sample is defrosted, manually mixed, and a representative sub-sample is taken

Fig. 7.1 Small sample cup from the FOSS DS2500 NIR scanner and a measuring cup containing the required amount of dried ground feed sample (~25 g) to adequately cover the bottom of the cup (Photo credit: Samantha Noel, Aarhus University, Denmark)

for freeze-drying. Importantly, if the pigs have had a diet shift, the faeces samples must be taken after the animal has had at least 8 days to adapt to the diet.

- The homogenised samples of feed or feed materials are freeze-dried or oven-dried in an air oven at 65 °C for 72 h.
- After drying, samples should equilibrate to room temperature and ambient humidity for at least 2 hours and then be ground to a particle size of 1 mm or 0.5 mm (passed through a 1 or 0.5 mm grid).
- The scan can then be performed after equilibrating to ambient humidity for 24 hours and samples can be stored for periods of up to a month at room temperature (20 °C) in airtight containers. It is preferential to scan the samples within 1 month after drying, but it is possible to use samples stored for longer periods of time (more than a year) if they are kept cold (−20 °C) in airtight containers.
- If samples have been stored for a period of more than 1 year before scanning and have visible moisture/ice inside the containers, they should be re-dried in a forced air oven at 60 °C for 24 h to remove the excess moisture. They can then be allowed to equilibrate to room temperature and ambient humidity for 2 h and are ready for scanning after 24 h.

7.5.2 Scanning Procedure

7.5.2.1 Machine Checks

The NIR scanner should be in a room with controlled temperature (preferably between 18 and 25 °C, but it is more important that the temperature is stable from day to day). The NIR machine should be turned on for a minimum of 30 minutes before scanning to allow the lamp to warm up. Once the machine is warmed up, open the machine software and allow it to run instrument diagnosis. A check sample (external reference standard) should be run, and if the reading falls in the expected range, scanning of the samples can proceed. If either the internal or external checks fail, the machine should be recalibrated.

7.5.2.2 Measurements

On the machine software, choose the calibration model you wish to use. The sample cup has a quartz bottom for scanning, and a notch in the sample cup means there is only one possible orientation to place it in the holder (Fig. 7.1). Check that the scanning cup and the surface of the instrument are clean before preparing the sample. The ground sample is scooped into a scanning cup so that a minimum of 1.5 cm covers the cup's surface (but preferably more) and no possibility of gaps exists. The sample is lightly tapped to remove air pockets but not pressed into the cup. The sample cup is placed into the scanner, whereby the NIR spectrum is recorded from the sample in reflectance mode. The scanner lid must be closed before scanning.

Once the scan is complete, the sample can be returned to its container, and the sample cup can be cleaned, to prepare for the next sample. Carefully remove all residue of the previous sample with a brush or compressed air. Periodically and at the end of all the measurements, perform a more thorough cleaning with a lint-free damp cloth of the sample cups and the surface of the instrument, where the cups are placed.

7.5.2.3 Data Acquisition

Data is recorded automatically by the instrument software (Fig. 7.2). Spectrum acquisition of the sample is dependent on the NIR scanner used. Some examples include:

- Bruker MPA (FT-NIR). Spectra are acquired between wavenumber 12,500 and 4000 cm^{-1} (wavelength from 800 to 2500 nm). The reading is done in reflectance mode in a quartz bottom sample cup (60 mm). Each sample receives an average of 64 scans.
- Foss NIRS DS2500 feed analyser (FOSS Analytical A/S, Silver Springs, MD, USA). Each scan was the average of 32 scans from various positions on the sample cup using the wavelength range from 400 to 2500 nm, with data recorded every 0.5 nm.

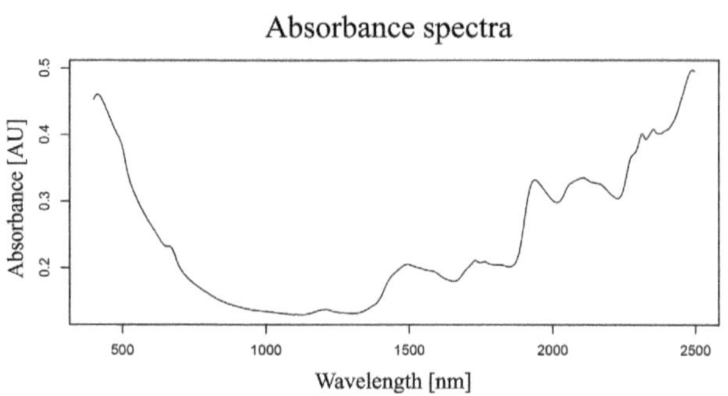

Fig. 7.2 Example absorbance spectra of a pig faecal sample

7.5.3 Prediction of Digestibility

The apparent ileal digestibility of amino acids is predicted from the sample scans of feed or faeces from the NIR machine software. Samples with parameters that are not within the range of the calibration equation (meaning that they are not a sample type that is used for generating the calibration, or the NIR spectra exceed the limits for Global H, Neighbourhood H, or T-statistics) should not be reported. The machine software will flag samples that exceed the spectra limits and give a warning for those nearing the spectra limit. Apparent digestibility results are reported as a percentage of nutrient intake digested in the ileum (so not recovered in the terminal ileum), without adjusting for endogenous losses (Stein et al. 2007).

7.5.4 Quality Control

To guarantee continued calibration accuracy over time, in addition to NIR quality checks, the calibration model should be annually assessed by measuring a subset of new samples (5–10% of the annual sample number) with the chemical reference methods and adjusting the calibrations if needed.

7.6 Compliance with the 3R Principles

Using NIR spectroscopy to predict digestibility aligns well with the 3Rs principles (replacement, reduction, and refinement) in animal research. NIR can replace and reduce the need for live animal testing by providing accurate predictions of feed digestibility and nutritional value. Predictions of apparent ileal amino acid digestibility based on feed samples replace the need for the use of animals altogether, and predictions based on faeces samples are non-invasive, requiring only the collection of faeces. By using NIR, you may also obtain predictions for multiple factors at once, reducing the number of analyses that need to be performed on a sample.

NIR spectroscopy is a non-invasive method that is fast and environmentally friendly. It refines the process of nutritional evaluation by providing fast, reliable estimates with the possibility of using the results in real-time applications like precision feeding. The collection of faeces or feed is also method refinement, as ileal digestibility is normally determined by more invasive canulation or slaughter of the pig.

7.7 Conclusions

The advantages of NIR analysis are that it is rapid, non-destructive, environmentally friendly and cost effective without needing invasive animal experimentation. The disadvantages are that it is an indirect method that requires the development of robust, well tested prediction models to give good results, and the results are only reliable if the samples fall into the range of variation of the samples that were used to generate the model.

Acknowledgements The authors would like to thank Professor Ruurd Zijlstra (University of Alberta, CA) and Dr. Karl Schedle (Boku University, AU) for reviewing and improving earlier versions of this SOP.

References

Baeten V, Pierna JAF, Lecler B et al (2016) Near infrared spectroscopy for food and feed: a mature technique. NIR News 27:4–6

Bastianelli D (2013) NIRS as a tool to assess digestibility in feeds and feedstuffs. In: Proceedings of the International Congress on Advancements in Poultry Production in the Middle East and African Countries, Antalya, Turkey, 21–25 October 2013, pp 1–21

Bastianelli D, Bonnal L, Jaguelin-Peyraud Y et al (2015) Predicting feed digestibility from NIRS analysis of pig faeces. Animal 9(5):781–786. https://doi.org/10.1017/s1751731114003097

Boisen S, Fernández JA (1997) Prediction of the total tract digestibility of energy in feedstuffs and pig diets by in vitro analyses. Anim Feed Sci Technol 68:277–286

Chen GL, Zhang B, Wu JG et al (2011) Nondestructive assessment of amino acid composition in rapeseed meal based on intact seeds by near-infrared reflectance spectroscopy. Anim Feed Sci Technol 165(1–2):111–119. https://doi.org/10.1016/j.anifeedsci.2011.02.004

Fontaine J, Schirmer B, Horr J (2002) Near-infrared reflectance spectroscopy (NIRS) enables the fast and accurate prediction of essential amino acid contents. 2. Results for wheat, barley, corn, triticale, wheat bran/middlings, rice bran, and sorghum. J Agric Food Chem 50(14):3902–3911

Knudsen KEB, Noel SJ, Jørgensen H (2023) Assessment of the nutritive value of individual feeds and diets by novel technologies. In: Kyriazakis I (ed) Smart livestock nutrition. Smart animal production, vol 1. Springer, Cham. https://doi.org/10.1007/978-3-031-22584-0_4

Millet S, Aluwe M, Van den Broeke A et al (2018) Review: pork production with maximal nitrogen efficiency. Animal 12(5):1060–1067. https://doi.org/10.1017/S1751731117002610

Noblet J, Perez JM (1993) Prediction of digestibility of nutrients and energy values of pig diets from chemical analysis. J Anim Sci 71:3389–3398

Noel SJ, Jorgensen HJH, Bach Knudsen KE (2021) Prediction of protein and amino acid composition and digestibility in individual feedstuffs and mixed diets for pigs using near-infrared spectroscopy. Anim Nutr 7(4):1242–1252. https://doi.org/10.1016/j.aninu.2021.07.004

Noel SJ, Jorgensen HJH, Knudsen KEB (2022) The use of near-infrared spectroscopy (NIRS) to determine the energy value of individual feedstuffs and mixed diets for pigs. Anim Feed Sci Technol 283:115156. https://doi.org/10.1016/j.anifeedsci.2021.115156

Patience J (2018) The theory and practice of feed formulation. In: Moughan PJ, Hendriks WH (eds) Feed evaluation science. Wageningen Academic Publishers, pp 457–490

Schiborra A, Bulang M, Berk A et al (2015) Using faecal near-infrared spectroscopy (FNIRS) to estimate nutrient digestibility and chemical composition of diets and faeces of growing pigs. Anim Feed Sci Technol 210:234–242. https://doi.org/10.1016/j.anifeedsci.2015.10.011

Stein HH, Fuller MF, Moughan PJ et al (2007) Definition of apparent, true, and standardized ileal digestibility of amino acids in pigs. Livest Sci 109(1–3):282–285. https://doi.org/10.1016/j.livsci.2007.01.019

Wang LF, Zijlstra RT (2018) Prediction of bioavailable nutrients and energy. In: Moughan PJ, Hendriks WH (eds) Feed evaluation science. Wageningen Academic Publishers, pp 337–386

Measurement of Salivary Cortisol by Enzyme Immunoassay

8

Liza R. Moscovice [iD] and Ulrike Gimsa

Abstract

There is increasing interest in measuring hormones in saliva as an alternative to the invasive and stressful procedures required to obtain blood samples from pigs. The hormone cortisol is released from the adrenals after activation of the hypothalamic-pituitary-adrenal (HPA) axis in response to arousing stimuli. There are a wide range of commercially available enzyme immunoassays (EIAs) that are suitable for measuring salivary cortisol (sCORT), making measurements in saliva both feasible and relatively cost-effective. However, greater attention to conducting and reporting analytical validations of the method in each lab is critical for increasing acceptance of non-invasive hormone measurements. Here, we describe the materials and procedures to validate and use commercially available EIAs for the measurement of sCORT in pigs. We focus on steps to prepare saliva samples for analyses and give advice on selecting appropriate EIA kits and additional preparation of materials and equipment to ensure proper analyses. We describe the gold standards for validations of hormone measurements by EIA, which include reporting of parallelism, precision and accuracy/recovery. We include supplementary materials with examples of calculations of the outcomes of different validation steps. This SOP thus provides thorough guidelines for practitioners to implement sCORT measurements in pigs in the laboratory via EIA.

Supplementary Information: The online version contains supplementary material available at https://doi.org/10.1007/978-3-032-24899-2_8.

L. R. Moscovice (✉) · U. Gimsa
Psychophysiology Working Group, Research Institute for Farm Animal Biology (FBN), Dummerstorf, Germany
e-mail: moscovice@fbn-dummerstorf.de

97

Keywords

Pigs · Laboratory methods · Analytical validation · Non-invasive

8.1   Introduction

Cortisol (ATOL_0002287) is a steroid hormone that is released peripherally from the adrenals after activation of the hypothalamic-pituitary-adrenal (HPA) axis (ATOL_0000946) in response to physically or psychologically arousing stimuli. Cortisol release triggers other physiological changes that help individuals to cope with challenges. While cortisol is often measured as an indicator of reactivity to negative and stressful events, cortisol can also be released in response to positive challenges, such as exercise or sexual stimuli, and cortisol release also has important functions in regulating daily circadian rhythms (ATOL_0000867) (McEwen et al. 2019). Peripheral cortisol concentrations in pigs can change within minutes in response to arousing stimuli and also fluctuate over hours and days, related to circadian rhythms. While blood (ATOL_0005631) sampling remains the primary method for measuring peripheral concentrations of cortisol and other hormones in pigs, there is increasing interest in identifying alternative methods for hormone measurement that do not require restraint and can minimise pain (ATOL_0000863) and distress to subjects (reviewed in Cerón et al. 2022). Measuring hormones in saliva presents several benefits over blood sampling. Saliva can be collected voluntarily from pigs in their home environments without restraint (Moscovice et al. 2022). As a result, samples can be collected by technicians and animal handlers on-site, without veterinary supervision. Well-designed studies measuring changes in salivary cortisol (sCORT) (ATOL_0005349) in response to short-term stimuli and also over longer time frames can provide insights into emotional states and physiological conditions in pigs. In addition to improving welfare (ATOL_0000765) by minimizing any pain or stress to pigs related to sampling procedures, measurement of cortisol in saliva can also benefit research, since sCORT is more likely than plasma cortisol (ATOL_0005348) to reflect changes in cortisol concentrations related to external variables of interest, rather than to the sampling procedure itself (e.g. Marcet Rius et al. 2018). However, methodologies to measure hormones in saliva have been developed relatively recently in comparison to methods for hormone measurement in plasma, and more efforts are needed to standardise methods for collecting (Chap. 4) and measuring hormones in saliva across research groups, to increase assay sensitivity and to reduce any potential interference from other background analytes that may be present in saliva (reviewed in Bellagambi et al. 2020). Related to this, consistency in reporting basic analytical validations of laboratory methods when presenting results is critical to increasing acceptance of non-invasive hormone measurements.

8.2   Goals and Scope of the Procedure

The goals of this SOP are to introduce a protocol for the measurement and validation of sCORT in pigs via enzyme immunoassay (EIA). In comparison with other
methods for hormone measurement, such as radio immunoassays (RIA), or liquid
chromatography-mass spectrometry (LC-MS), EIAs are relatively cheap and easy
to implement in labs, and several commercially available EIAs have already been
validated for the measurement of sCORT in pigs (e.g. Goursot et al. 2019; Moscovice
et al. 2022). Nonetheless, every research group needs to initially validate their chosen EIA in their own lab with pooled test samples from pigs, prior to embarking on
analyses of research samples. You should validate your assay method the first time
that you analyse a hormone in a new matrix (e.g. saliva) and/or in a new species. In
addition, when you measure hormones in a new breed or age cohort of pigs, it may
be necessary to confirm the most appropriate dilution factor for the samples (see
below). Studies that do not adhere to internationally recognised standards for reporting laboratory validations lack quality assurance (Taverniers et al. 2004) and can
slow the progression of research. Here, we describe the materials and procedures to
validate and use commercially available EIAs specifically for the measurement of
sCORT in pigs. For additional information on recommended materials and procedures for collecting saliva samples from pigs, please refer to our related SOP on
saliva sampling in pigs (Chap. 4).

8.3   Materials and Equipment

8.3.1   Enzyme Immunoassay Kit and Reagents

For best results, we recommend using a commercially available EIA kit that has
been previously validated for the measurement of sCORT in humans. Full validations of saliva measurement in humans should be reported with the kit's informa-
tional materials. One option is the 'Cortisol free in Saliva ELISA kit' from
Demeditec Diagnostics Gmbh (Kiel, Germany), which has been validated in humans
and has been further validated by our lab for the measurement of salivary cortisol in
pigs (Goursot et al. 2019; Moscovice et al. 2022). Other kits may also be used.

> **Important**
> Note that a validation of the kit for use with saliva by the manufacturer is
> important to assess the kit quality (which can vary greatly), but this does *not*
> preclude conducting your own internal validation of measurement of salivary
> cortisol in pigs in your lab. In addition, many cortisol assay kits are designed
> and validated for measuring cortisol in human plasma, which is one of the
> reasons why the additional validations described in Sect. 8.5.1 are also critical
> when measuring sCORT.

You may need to purchase additional cortisol standard and assay buffer from your ELISA kit manufacturer for your validation steps. Most companies allow you to purchase additional kit standards and assay buffer separately from the kits if needed.

8.3.2 Additional Lab Equipment

- 4 °C refrigerator.
- Vortex mixer.
- Centrifuge.
- Plate shaker with incubator (to maintain 25 °C).
- Calibrated microtiter plate reader (plate will be read at 450 ± 10 nm).
- Calibrated variable precision micropipettes (50 µL, 100 µL, 200 µL).
- Distilled or deionised water.
- Eppendorf tubes (0.1 to 0.5 mL tubes are sufficient for test samples. Larger tubes may be needed for validation steps.)

8.4   Prerequisites and Preparation

8.4.1   Preparation of a Pooled Saliva Sample
for Assay Validations

For analytical validations of your assay, you will need a pooled sample of 3–4 mL saliva from several different pigs. Your sample should be from multiple animals to ensure that your pool is representative of your subjects. Ideally, you should use samples that are expected to have a relatively high concentration of cortisol (e.g. samples collected after a known biological stressor), so that cortisol concentrations can be detected in serial dilutions of the pooled sample, ranging from high to low concentrations. To make your pooled sample, thaw samples to room temperature (if frozen), combine all samples in one large tube and vortex for 5–10 secs. Centrifuge at 3500 x g for 5 min (you can aliquot the mixed sample into smaller tubes for centrifuging if necessary). Combine the supernatant from all of the tubes into one new, larger tube. When combining samples, pipette carefully to make sure that you do not include any pellet from the bottom of any tubes, as this may introduce interference to the measurements. Vortex the pooled sample for 5–10 secs. Either use the pooled sample immediately, or store it for the short-term (a few days) at 4 °C, or for longer-term at −20 °C. You may want to store the pooled sample in multiple smaller aliquots (e.g. 1 mL) for ease of later use.

8.4.2 Preparation of High- and Low-Cortisol Concentration Control Samples

Coefficients of variation (CVs) are a measure of the variance in replicates of the same sample, either run on the same plate (intra-assay CVs) or on different plates (inter-assay CVs). They are an important measure of consistency in your assay standards and measurements across plates. To calculate CVs, you should run control samples with known cortisol concentrations on every plate. If you have enough reserve saliva, you can make high- and low-concentration controls by combining several samples that were previously measured in pigs and found to have high (~ 10–20 ng/mL) or low (~ 1–3 ng/mL) cortisol concentrations, respectively. If you do not have any reserve saliva samples from pigs, you can create a high-concentration control by using an ~10 ng/mL concentration of cortisol standard from an ELISA kit. For low-concentration controls, we recommend using an ~1 ng/mL cortisol standard. It is recommended to create your high- and low-concentration controls in advance of running your EIA, by making multiple 250 µL aliquots from larger pools, and freezing them until you prepare your plate. Aliquots of 250 µL of high- and 250 µL of low-concentration pools are sufficient to obtain the 200 µL of each pool required to run each control twice on each plate.

8.4.3 Sample and Reagent Preparation

Samples with a minimum of 0.1 mL saliva can be analysed for cortisol in duplicate. Analysis of samples in duplicate is critical for accurate assay measurement. Samples can be frozen after collection until measurement. In preparation for measurement, frozen samples should be thawed to room temperature. Samples should be spun at 2300 x g for 5 min. During this time, new Eppendorf tubes can be pre-labelled, and in case any dilution is needed, an appropriate amount of assay buffer can be added to each tube to obtain the desired dilution factor. For pigs, we recommend combining 75 µL assay buffer with 75 µL of sample, for a 1:2 dilution factor (1 part sample in 1 part buffer). You can also determine the ideal dilution factor for your test population by measuring different pooled samples at different dilution factors and identifying a dilution factor with which the majority of samples fall within the kit standard curve. If there is >0.1 mL saliva available per sample, this will be useful in case samples need to be re-run. Repeat measurements will be needed in case any samples measure above the standard curve, in which case they should be further diluted and measured again. Samples that measure below the standard curve should be run a second time without any dilution. If diluting your samples, add the assay buffer and then vortex each sample briefly to mix. Typically, n = 38 test samples can be run in duplicate on one plate, plus four control samples (described below) and additional internal plate controls.

Read the assay kit instructions carefully for detailed information on reagent preparation. Usually, kit reagents should be stored at 4 °C and taken from the refrigerator and warmed to room temperature (e.g. 18–25 °C) before using. This can take between 30 min and 1 h, so make sure to allocate time for this step if needed.

8.4.4 Preparation of Microtiter Plate Reader

You will initially have to set up a program on your plate reader for the measurement of cortisol. With the plate reader software, you should be able to indicate the placement and concentrations of the standard curve, the placement of various internal controls and also the placement and abbreviation for each sample that you are running on a plate. If you performed dilutions on your test samples, you can indicate the dilution factor on your plate layout in order to obtain modified concentrations that take into account the dilution factor. Otherwise, you will have to manually calculate the correct concentrations later, accounting for any dilutions. You should indicate a wavelength of 450 nm for measuring cortisol (this should also be described in your EIA kit protocol).

8.5   Description of the Procedure

8.5.1 Assay Validation Steps

8.5.1.1 Parallelism

Parallelism tests whether the concentration response of cortisol in the matrix (in this case, saliva) is similar to the response of cortisol in the assay buffer and is determined by making a serial dilution of pooled saliva samples from your subjects and comparing the resulting concentrations to the standard curve of your assay. Serial dilutions that are parallel to the standard curve indicate that cortisol can be accurately measured in the matrix. In addition, parallelism can help to determine the ideal dilution factor to use for your subjects. In general, for young pigs, we have found that a dilution factor of 1:2 is able to capture 98% of the sample concentrations within the standard curve. However, the ideal dilution factor may differ with breed, age, etc., so you should also confirm this through your own dilution curve. To perform a parallelism, you will need to use approximately 0.4 mL of a preferably high-concentration pooled saliva sample (Sect. 8.4). Label five empty Eppendorf tubes from 1 to 5. Add 200 µL assay buffer from the cortisol kit to each tube. Add 200 µL of your pooled sample to tube 1. Vortex tube 1 well. Then add 200 µL of the mixed sample from tube 1 to tube 2. Vortex tube 2 well and then add 200 µL of the sample from tube 2 to tube 3. Repeat these steps for tubes 3–5. After this, you will have a serial dilution of 1:2, 1:4, 1:8, 1:16 and 1:32 concentrations of your original pooled sample in assay buffer. You should load 100 µL of the original pooled sample and 100 µL of each of the dilutions (6 samples in total, each run in duplicate) on the EIA plate, in order of their dilution factors. Note that other serial dilutions are also possible; this is just one example. If you are concerned that your pooled sample may not be high enough in concentration to detect cortisol in all of the dilutions, you can use a different serial dilution factor, or you can spike the pooled sample with a known amount of the cortisol standard from your kit (e.g. by adding 25 µL of a ~ 30 ng/mL standard to your pooled sample). This will increase the concentration of cortisol and can help to ensure that all of the dilutions are within the standard curve.

8.5.1.2 Precision

Assay precision indicates how accurately your method provides the same result when replicates of the same sample are measured repeatedly. Precision is based on the variability surrounding multiple measurements of high- and low-concentration control samples (Sect. 8.4). These high- and low-concentration controls can be prepared in advance and should be run in duplicate on each plate to calculate inter-assay coefficients of variation, or CVs. We recommend that you run one high-control and one low-control sample (in duplicate wells) in an early position on the plate (e.g. on columns 3 and 4) and run the other identical high- and low-control samples (in duplicate wells) at a later position on the plate (e.g. in columns 9 and 10). This will help to check for any effects of plate-shift related to the EIA procedure, which can occur if, for example, one of the pipetting steps takes too long to complete from the beginning to the end of the plate. In addition to running high- and low-concentration control samples on every plate to calculate inter-assay CVs, between 6 and 8 replicates of the high- and low-concentration controls should be run once on the same plate to measure intra-assay CVs. For this, you will need additional volumes of your high- and low-concentration controls, from which you will load 100 µL of each control (in duplicate wells) between 6 and 8 times on one plate.

8.5.1.3 Accuracy/Recovery

Recovery is your ability to measure the amount of a known analyte spiked into and recovered from a biological sample compared to the non-spiked sample. To determine recovery, you will need to use two to three different pools of saliva samples, each of approximately 0.5 mL. The goal is to create multiple pools that each have a different starting concentration (ideally from relatively low to relatively high concentrations). You will then create high-, medium- and low-concentration spiked samples from each pool, by spiking each pool with cortisol standards from the kit. Label three Eppendorf tubes as follows: Pool1-H, Pool1-M, Pool1-L. To create the recovery samples for pool 1, add 100 µL of pooled sample 1 to each of the three tubes. Spike the Pool1-H with 100 µL of a 30 ng/mL cortisol standard from the cortisol kit. Spike Pool1-M with 100 µL of a 7 ng/mL cortisol standard from the kit. Spike Pool1-L with 100 µL of a 0.1 ng/mL standard from the kit. You will then have high-, medium- and low-concentration spikes of pool 1. Perform the same steps with the second and third pools, to achieve a high-, medium- and low- concentration pool 2 and pool 3. Run all of the spiked samples, plus the original unspiked pools, on your plate next to each other. Note that this is just one example of how to create spiked samples for recovery. Other concentrations of cortisol standard can also be used to create different ranges of high-, medium- and low- concentration spiked samples. The goal should be to create a range of concentrations of different pooled samples that span the standard curve of your assay.

8.5.1.4 Validating your Method for Salivary Cortisol Measurement

We recommend that you run your validation samples at an early stage, before you run most of your research samples (although running some initial research samples may be necessary to identify appropriate high- and low-concentration samples to use when making various controls). Related to this, and with the validation steps in mind, it may be useful to collect several mLs of saliva from several of your test subjects before collecting any research samples, for use in the validation steps. This can be achieved while habituating your test subjects to sample collection (Chap. 4). After you have validated your assay for measuring sCORT in pigs and confirmed the appropriate dilution factor for your samples, you can then run additional samples on subsequent plates. Remember to run duplicate high- and low-control cortisol standards on every plate. Your parallelism, recovery and intra-assay precision only need to be run once if the results are within acceptable ranges (e.g. CVs < 20%, recoveries between 80 and 120% and no significant differences in slopes between serial dilutions of matrix and serial dilutions of kit standard (see Andreasson et al. 2015). If results of any of the validation steps are poor, the problematic validation(s) should be repeated until results are within acceptable ranges. This will ensure greater validity of your subsequent measurements of research samples.

8.5.2   Measuring Cortisol Using an Enzyme Immunoassay

8.5.2.1 Preparation and Measurement of Samples on an EIA Plate

Once you have prepared the research samples with proper dilutions (Sect. 8.4.3) and have prepared your various samples for validations as needed (Sect. 8.5.1), you should now refer to the assay kit instructions for additional important information for running your plate. Note that this protocol does not present complete instructions for assay measurement, which can vary from kit to kit. You should find detailed instructions on how to prepare a standard curve and other necessary internal controls for the plate. Follow the assay kit instructions closely for these steps. Once you have prepared the kit standard curve and any other internal controls as described, then pipette your validation samples, research samples and the other kit control samples onto the EIA plate, following the kit instructions. Your kit will also describe one or more incubation steps prior to the measurement of your samples. Again, follow the kit instructions closely regarding the timing and conditions (related to light exposure, temperature, with or without shaking, etc.) for the incubation steps. You should also program a plate reader in advance with the relevant information for your cortisol plate (Sect. 8.4.4). After the final incubation step, place your plate in the plate reader and run the pre-designated program for sCORT to obtain your results (Fig. 8.1). Salivary cortisol concentrations will usually be given in ng per mL.

8.5.2.2 Reporting your Results

When reporting your cortisol results, you should also include the results of your validations. For parallelism, you may want to include a figure comparing the serial dilutions of pooled salivary cortisol and the standard curve (see Supplementary

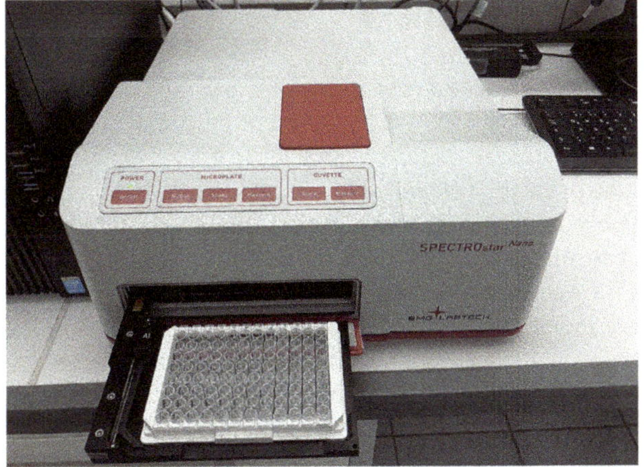

Fig. 8.1 Example of a microtiter plate reader, with an EIA plate being loaded for analysis. Photo credit: Liza R. Moscovice, FBN

Worksheet 8.1). You can also use a t-test to determine whether the slopes of the lines from the serial dilutions of pooled saliva samples and the standard curve differ significantly from each other (they should not). For recovery, you should report the average and standard deviation of recovery across your pooled, spiked samples (see Supplementary Worksheet 8.2). A recovery of 90–110% is considered to indicate a high degree of reliability. For precision, report the coefficients of variation for high- and low- concentration cortisol control samples across multiple plates, and for the high- and low- concentration cortisol samples measured multiple times on the same plate. Coefficients of variation are calculated by taking the standard deviation of a set of measurements divided by the mean of that set of measurements (see Supplementary Worksheet 8.3). In general, inter- and intra-assay CVs for sCORT measurements at or below 10% indicate a high degree of precision. Refer to the ESM worksheets for examples of how to calculate CVs, parallelism and recovery.

8.6   Compliance with the 3R Principles

This protocol complies with the 3R principles by improving the validity of non-invasive measurements of sCORT in pigs. Salivary cortisol can be collected from pigs in their home environment, without any restraint or pain, and with minimal disruption to the pig's ongoing behaviour, as long as pigs are habituated in advance to the procedure (Chap. 4). Increased standardization and validation of methods for non-invasive sampling of hormones will improve animal welfare, by increasing acceptance and utilization of methods that reduce pain and distress associated with sample collection (MacArthur Clark 2018). We put special emphasis here on validation steps for sCORT measurements, since we believe that presenting rigorous validations of this non-invasive method will lead to wider acceptance of this method as a viable alternative to measurements of cortisol in plasma.

8.7   Conclusions

Here we present a protocol for measuring cortisol concentrations in saliva samples from pigs via EIA. We also present gold standards for validating hormone measurements, which should be conducted whenever measuring hormones in a species or matrix for the first time (Andreasson et al. 2015). Enzyme immunoassays are a relatively inexpensive and widely available method to measure steroid hormones like cortisol, and results from EIAs are reliable when the kit instructions are closely followed, and when the additional controls and validations specified in this chapter are included. The disadvantages of using EIAs are that kits can sometimes vary in their quality and not all kits are previously validated by the company for the measurement of cortisol in saliva. Thus, care should be taken to select a kit from a reputable company, and ideally one that has been previously validated by that company for the measurement of salivary cortisol in humans or other species. It is important to consider that while reliable, inexpensive and relatively easy to run, EIAs are also less accurate than other more expensive methods such as liquid chromatography-mass spectrometry (LC-MS), which has the advantage of being able to measure multiple glucocorticoid stress hormones, such as cortisol and corticosterone, along with other steroid hormones, in smaller volumes of saliva samples and with greater accuracy compared to EIAs. Thus, decisions about the use of EIAs versus other methods for measuring sCORT should be based on your specific research goals and budget.

Acknowledgments We are grateful to Winfried Otten, Research Institute for Farm Animal Biology (FBN) and Elodie Merlot, National Research Institute for Agriculture, Food and Environment (INRAE), for their valuable feedback on earlier versions of this chapter.

References

Andreasson, U., Perret-Liaudet, A., van Waalwijk van Doorn, L. J. C., et al (2015). A practical guide to immunoassay method validation. Front Neurol, 6, 179. doi:https://doi.org/10.3389/fneur.2015.00179

Bellagambi FG, Lomonaco T, Salvo P et al (2020) Saliva sampling: methods and devices. An overview. Trends Anal Chem 124:115781. https://doi.org/10.1016/j.trac.2019.115781

Cerón JJ, Contreras-Aguilar MD, Escribano D et al (2022) Basics for the potential use of saliva to evaluate stress, inflammation, immune system, and redox homeostasis in pigs. BMC Vet Res 18:81. https://doi.org/10.1186/s12917-022-03176-w

Goursot C, Düpjan S, Kanitz E et al (2019) Assessing animal individuality: links between personality and laterality in pigs. Curr Zool 65(5):541–551. https://doi.org/10.1093/cz/zoy071

MacArthur Clark J (2018) The 3Rs in research: a contemporary approach to replacement, reduction and refinement. Br J Nutr 120(s1):S1–S7. https://doi.org/10.1017/S0007114517002227

Marcet Rius M, Cozzi A, Bienboire-Frosini C et al (2018) Selection of putative indicators of positive emotions triggered by object and social play in mini-pigs. Appl Anim Behav Sci 202:13–19. https://doi.org/10.1016/j.applanim.2018.02.002

McEwen BS (2019) What is the confusion with cortisol? Chronic Stress 3:2470547019833647. https://doi.org/10.1177/2470547019833647

Moscovice LR, Gimsa U, Otten W et al (2022) Salivary cortisol, but not oxytocin, varies with social challenges in domestic pigs: implications for measuring emotions. Front Behav Neurosci 16:899397. https://doi.org/10.3389/fnbeh.2022.899397

Taverniers I, De Loose M, Van Bockstaele E (2004) Trends in quality in the analytical laboratory. II. Analytical method validation and quality assurance. TrAC - Trends Anal Chem 23(8):535–552. https://doi.org/10.1016/j.trac.2004.04.001

Measuring Peripheral Temperature Using Infrared Thermography

Helena Telkänranta

Abstract

Infrared thermography, also known as thermography or thermal imaging, is a technology for measuring the distribution of temperature across a given surface. In animal health and welfare, the development of thermographic methods for research is still in its early infancy. When used correctly, thermography provides a non-invasive tool for acquiring information on physiological processes that affect the temperature on some part of an animal's surface: e.g. thermoregulation, local inflammation and functioning of the autonomic nervous system. The protocol presented here will describe the technical steps needed for reliable thermographic data collection while preventing adverse effects on animal welfare, including the following: (1) selecting the correct equipment and software; (2) identifying the correct time frame and locations for thermal recordings; (3) selecting between recording single images or thermal video, with or without numerical data; (4) ensuring the safety of the pigs and the thermal camera; (5) habituating the pigs to thermal recordings; (6) identifying individual pigs in thermal recordings; and (7) setting the correct parameters in the thermal camera before and during the recording sessions. Users are also advised to gain a thorough understanding of the thermal biology of the physiological process they will be studying.

Keywords

Pigs · Thermography · Thermal imaging · IRT · Non-invasive

H. Telkänranta (✉)
Arador Innovations Ltd, Helsinki, Finland
e-mail: helena.telkanranta@arador.fi

© The Author(s) 2026
L. R. Moscovice et al. (eds.), *Standard Operating Procedures for Better Pig Research (SOPig)*, https://doi.org/10.1007/978-3-032-24899-2_9

9.1   Introduction

Infrared thermography (IRT), also known as thermography or thermal imaging, is a technology for measuring temperature distribution across a given surface at a distance. Any surface will naturally emit occasional photons, and the intensity of the photon stream depends on the temperature of the surface. Because of the wave-particle dualism effect in particle physics, this photon stream exists simultaneously as a wave of radiation, and therefore, the intensity of the photon stream can also be expressed as the wavelength of the corresponding radiation. A thermal camera contains an array of sensors detecting the wavelength of that part of the radiation which falls within a certain bracket in the infrared range. Each sensor stores this information as a data point. This information is usually visualised as a thermal image, sometimes called a thermogram. Each pixel represents the wavelength detected by that sensor, transformed to a data point expressed as a value of temperature. For the purposes of either visual monitoring or finding the pixels corresponding to the area of interest (e.g. a given body part) in numerical data collection, thermal images are often presented as false-colour heat maps, in which different colours symbolise different temperature ranges (Fig. 9.1).

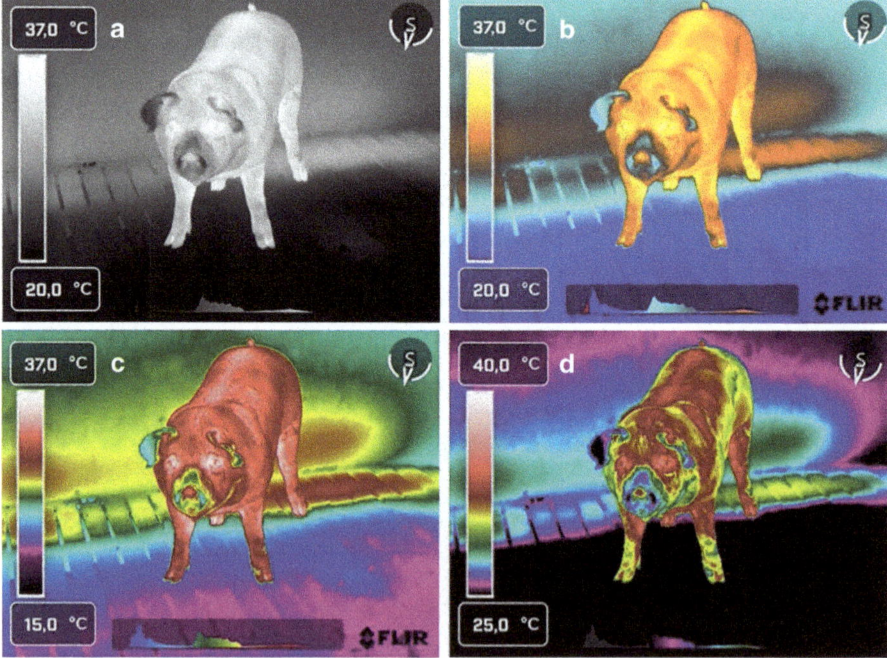

Fig. 9.1 Different visual representations of the same thermal image, containing the same numerical data. (**a**) A greyscale image with a scale on the left. Surface temperatures below that range are shown in black, and those above it are shown in white. (**b**) The image with colours added to highlight temperatures above and below average. (**c** and **d**) Two versions of the image made with the same colour palette ('rainbow high contrast'); the differences are in the temperature range shown as colours and in the colour distribution: linear (**c**) vs. non-linear (**d**)

Important

In addition to the actual surface being imaged, there are several other factors affecting the wavelength of the incoming radiation, such as how much the surface to be measured deviates from being parallel to the camera lens (the further the surface is from being parallel, the more photons will end up elsewhere and not on the camera lens), whether there are additional sources of infrared radiation (such as sunshine, an infrared lamp or other nearby heating devices) and other factors. The data on a thermal image is a combination of the actual temperature distribution across the surface of interest, plus the effects of external factors. Planning thermal imaging studies therefore requires a good understanding of thermal physics to minimise confounding factors (Minkina and Dudzik 2009).

Outside of the biological sciences, IRT has had several practical uses for decades, such as detecting where heat is leaking from buildings, finding people in a landscape in rescue or military settings, detecting faulty electrical components in a machine based on their elevated temperature, monitoring industrial combustion processes, etc. In each of these uses, it is sufficient to visually inspect a thermal image and detect an object whose temperature differs from its environment, without a need for precise data. The same is true of some medical and veterinary uses, in which the goal is to find inflammation, or other temperature-altering processes, such as a tumour, that are near enough to the skin (ATOL_0005636) to cause surface temperature effects (Rekant et al. 2016). Similarly, IRT can be used to detect substantial problems with thermoregulation (ATOL_0000857), if these have prominent effects on surface temperature. One such example is the detection of hypothermic young piglets (Kammersgaard et al. 2013). A much more challenging field, still in its infancy, is the use of thermal imaging to monitor the functioning of the autonomic nervous system (ANS) and measure some physiological correlates of animal health and welfare (Franchi et al. 2024; Ghezzi et al. 2024). Development of research methods in this field requires not only a good understanding of thermal physics but also a good understanding of the physiology of the species, to distinguish the effects of interest from those of other physiological processes, such as exercise, metabolism and the sleep-wake cycle, which often affect surface temperatures of the same body parts as ANS activity does. If sufficient interdisciplinary knowledge is utilised in the development process, there is considerable potential for future development of new IRT methods to measure animal health and welfare. While there are research groups working to develop such methods, at present, there are no validated IRT methods yet to measure states, such as positive vs. negative emotions, or welfare as a long-term state. It is also useful to note that, contrary to common belief, measuring fever with thermography can easily cause false negatives if applied with insufficient physiological knowledge. The skin is a major thermoregulatory organ, and while the body is in heat-conservation mode during increasing fever, the vasoconstriction in skin vasculature may keep the animal's surface temperature at a normal or, especially in peripheral body parts, even below-normal range. What can be measured at present are physiological processes as such, whenever they have a substantial enough effect on surface temperature distribution, such as thermoregulation, local inflammation and functioning of the autonomic nervous system.

9.2 Goal and Scope of the Procedure

This protocol describes the steps needed for planning a technically reliable thermal imaging session and data collection including the following: (1) selecting the correct type of thermal camera, lens, other equipment and software; (2) identifying the correct time frame and locations for thermal recordings; (3) selecting between recording single images or thermal video, and with or without numerical data; (4) ensuring the safety of the pigs and the thermal camera; (5) habituating the pigs to thermal recordings; (6) identifying individual pigs in thermal recordings; and (7) setting the correct parameters in the thermal camera before and during the recording sessions. Knowledge of how to extract temperature data from thermal images or videos by using analysis software will not be covered here, but there are tutorials available on the webpages of the manufacturers of thermal cameras, explaining the basics of using the software. Details of data extraction will depend on the research question, and there is no comprehensive manual on that as yet. This protocol can be used with pigs of any age and sex, regardless of whether they are housed in a social group or not, and in any environment where there are no external physical confounders such as infrared heating lamps or direct sunlight. The experimental unit in most cases will be the individual, with the exception of those veterinary settings where the research question concerns temperature differences within an individual, e.g. comparing a leg joint with inflammation to a healthy joint in another leg. This protocol is based on the assumption that the users already have a thorough understanding of the thermal biology of the physiological process they will be studying, as well as of the other processes that affect the surface temperature of the same body parts.

9.3 Materials and Equipment

9.3.1 Thermal Camera and Lens

Selecting the right type of thermal camera depends on three factors: (1) whether the users will need numerical data, which is the case in most scientific research involving thermography, or whether they plan to do visual inspection only, which is the case in some practical uses for monitoring thermoregulation or local inflammation; (2) how high a temporal resolution is needed, i.e. how many thermal images per unit of time will be needed; and (3) how high a spatial resolution is needed, i.e. how many pixels the thermal images will have to include.

1) To obtain numerical data, it is necessary to select a camera model that can produce radiometric thermal images, i.e. images in which the temperature value linked to each pixel is stored and accessible in the image files; or radiometric video, which is a sequence of radiometric images.
2) The temporal resolution needed will depend on whether the pigs may move during imaging. For animal welfare (ATOL_0000765) reasons, restraining pigs for the sake of an experiment should be avoided. If the experiment requires the pigs to stay still during imaging, positive reinforcement training is recommended. If the pig stays still during imaging, it is sufficient to have a thermal camera that captures radiometric images but not radiometric video. If the pigs will move during imaging, which is usually the case, it is necessary to use a thermal camera that can also capture radiometric video, preferably at a frequency of 30 frames per second (fps).
3) The spatial resolution needed will depend on how large the region of interest is (e.g. the tip of the tail vs. the entire back region) and how far from the thermal camera the pigs will be. If numerical data is needed, it is recommended that the region of interest will be represented by 10x10 pixels at a minimum, as data collection with the most common types of data collection software may otherwise give inaccurate readings at the borders of the area. If the thermal images will be used for visual inspection only, it is recommended that the area of interest is represented by 3x3 pixels at a minimum, to avoid random errors in single pixels.

Many thermal camera models include the option of changing the lens to acquire a wider or narrower field of view. Even when buying or renting such a thermal camera with one lens only, there is often the option of choosing the lens. Wide-angle lenses, with a field of vision (FOV) of 24 or more (which can also be expressed as a focal length (f) of 18 or less), are suitable for most indoor settings, where the pigs are near the thermal camera, within a few metres. Narrower lenses, such as FOV of 18 or less (focal length (f) of 29 or more), are suitable for most of the situations, where the pigs are further away from the camera.

There are several manufacturers of thermal cameras. The two manufacturers that sell the most thermal cameras are Teledyne FLIR and HIKMICRO. The US-based company Teledyne FLIR, which was known as FLIR before its fusion with Teledyne and which still uses FLIR as the trademark on its cameras, has the largest selection of different thermal cameras and types of analysis software and used to be the market leader by a substantial margin. The China-based company HIKMICRO has grown rapidly in recent years by manufacturing thermal cameras and software similar to some of the FLIR models, at a lower cost. For those considering purchasing a HIKMICRO camera, it is worth noting that using them for numerical data collection requires downloading Chinese software, which may be prohibited by some research institutes and other organisations.

In most research uses with pigs, the recommended type of thermal camera is that of FLIR T-series or similar, with a wide-angle lens of FOV 24 or higher, which is capable of recording radiometric video at 30 fps with a resolution of 640x480 pixels or higher. If the recording involves non-moving animals only, such as monitoring thermoregulation in suckling piglets or screening for joint inflammation or tail tip inflammation in pigs standing at the feed trough, a much lower-cost thermal camera is sufficient, such as one of the lower-end models in FLIR E-series or similar.

9.3.2 Equipment for Ambient Measurements, Data Storage, Camera Protection, Etc.

- Thermometer and hygrometer to measure ambient temperature and air humidity.
- A laptop for image analysis.
- A notepad and pen, or some other note-taking system such as voice recording on the phone, to keep a log as to which individual animal is in which thermal image or video.
- SD cards, if recording single thermal images only, or if recording radiometric video with a thermal camera capable of storing radiometric video on an SD card.
- If recording radiometric video with a thermal camera that requires streaming to an external storage device: an external hard drive, and either a USB cable, a Wi-Fi connection or a Bluetooth connection (note that Wi-Fi and Bluetooth are not supported by some thermal camera models) for connecting the thermal camera to the laptop.
- A carrying case for the thermal camera to protect it when not in use.
- A lens cap to protect the camera lens; this normally comes with the camera, but it is recommended to have a spare one in case it gets lost.
- Isopropyl alcohol (IPA) and cotton pads for cleaning the camera lens in case of staining.
- Optional: tripod for the thermal camera if all the thermal recordings are to be carried out in one or a few standard locations.

9.3.3 Software

Manufacturers of thermal cameras provide software for the processing and analysis of thermal images and videos. Depending on the manufacturer, some or all of the software products can be downloaded for free at the manufacturer's website. Advanced software products, intended for research use, usually require buying a licence.

Software will be needed for the following purposes: recording radiometric video on an external storage device, such as a laptop or an external hard drive; changing the colour palette in the thermal images or video after recording; changing the temperature ranges represented by each colour zone in the thermal images or video after recording; extracting and saving single thermal images from thermal videos; and extracting numerical data from thermal images.

Some research groups have developed their own bespoke software for data extraction from thermal images. This will save time at the two most time-consuming stages of data analysis: extracting numerical data from thermal images and exporting that data to a file format such as Excel for statistical analysis. Development of such bespoke software is especially useful with large quantities of thermal video, but it will require advanced programming skills.

9.4 Prerequisites and Preparation

9.4.1 Identifying the Time Frame for Thermal Recordings

When planning the timing of thermal recordings, the following aspects should be taken into consideration: (1) balancing out random fluctuations in the measurements of the thermal camera; (2) if recording a process involving temperature dynamics over time, including the entire relevant period; and (3) if comparing recordings from different days, standardising the time of day.

1) Random fluctuations in the measurements of the thermal camera occur in all uncooled thermal cameras with microbolometer sensors (referring to the type of thermal sensor used to detect IR radiation); all the normally used thermal cameras in research and monitoring are of this type. There are two ways to cancel this out: either by recording multiple images and taking the average of the temperatures of the region of interest in each image or by collecting the data in terms of within-image differences only, such as the temperature of the region of interest as compared to the temperature of a standard reference point.
2) When studying physiological processes that involve dynamics over time, it is usually necessary to plan the time frame so that it will include all of the relevant phases in that sequence of events. For example, when studying the functioning of the ANS, it is usually necessary to include first a baseline situation, then the activation of the sympathetic nervous system and then the subsequent downregulating effect of the parasympathetic nervous system until returning to the baseline. The baseline will require a situation in which pigs are not anticipating any forthcoming event. In many cases, hearing people approach the room before opening the door will already have triggered a sympathetic response, so planning is required to determine a way to get a true baseline, for example, by a researcher with a thermal camera spending enough time with well-habituated pigs. Once the stimulus to elicit a sympathetic response has occurred, the dynamics of sympathetic activation and parasympathetic downregulation will partly depend on the intensity of the response. There is limited research on pigs, but studies on cattle have shown the sympathetic response to reach its peak within a few minutes after the stimulus (often 1–2 min), and the return to baseline can vary significantly, in many cases taking more than 20 min (Stewart et al. 2010).
3) Diurnal rhythm will affect the surface temperature via several mechanisms, including variation in core body temperature and thermoregulation (whether the body is in heat-conserving, heat-dissipating or balanced mode). Thermoregulation will have a much more prominent effect than core body temperature on an animal's surface temperature.

9.4.2 Identifying the Locations for Thermal Recordings

When planning the location(s) for the thermal camera, the following aspects should be taken into consideration: (1) visibility of the region of interest to the thermal camera, (2) prevention of environmental confounders, (3) the angle between the camera lens and the surface to be measured and (4) the distance of the animal from the thermal camera.

1) The visibility of the region of interest to the thermal camera will involve planning how to prevent it from being obscured behind pen structures and/or behind other pigs; how to create a situation in which the pig's own body orientation presents the region of interest to the camera; and how to ensure the region of interest does not include pieces of straw, mud, ocular secretions or other substrates obscuring the animal's surface on parts of the region of interest.
2) Preventing physical confounders, i.e. those environmental factors that will increase or reduce the amount of infrared radiation emitted from the region of interest. Some of the most common examples include wet or moist areas on the animal's surface, infrared lamps, radiators and other heating devices, sunlight through a window, direct sunlight outdoors, wind and other strong air currents and nearby cold surfaces, such as a cold floor or wall in the winter.
3) Optimally, the surface to be measured should be parallel to the lens of the thermal camera. The more deviation from this there is, the more photons emitted by the surface end up in other directions and not the camera lens, making the surface appear cooler to the camera than it actually is. For flat surfaces, a deviation of less than 30 degrees is typically needed to avoid the measurement error from reaching statistical significance, but the actual safe limit in each case also depends on the magnitude of the effect. For curved surfaces, the safe limit is expected to be less than that, but there is no published research as yet on the threshold values for different curvatures.
4) The distance of the animal from the thermal camera is one of the parameters that will need to be set in the thermal camera before imaging. Measuring the dimensions of the pens or other facilities beforehand will help in estimating the animals' distances during imaging.

9.4.3 Selecting Between Radiometric vs. Non-radiometric Recording, and Between Single Images vs. Video

A sequence of thermal images in rapid succession, such as 8 or 30 frames per second, is called a thermal video. Thermal images and thermal videos can be either radiometric, i.e. containing the original numerical data in connection to each pixel, or non-radiometric, showing a heat map but retaining no numerical data. Radiometric images and videos are used in most research and monitoring purposes, because the existence of stored data in each pixel makes it possible to collect numerical data and to adjust the colour palette afterwards, bringing out more detail for visible inspection or highlighting specific temperature ranges for the purposes of communication. Collecting numerical data and adjusting the colours afterwards requires specialist software. Radiometric images can be viewed as.jpg files without specialist software, but radiometric video is compressed in a format (such as.csq or.seq) that can only be viewed and processed with software. Such software is available from the same manufacturers as thermal cameras. Non-radiometric thermal images are rare nowadays, appearing only in the very lowest-cost thermal cameras. Non-radiometric video is more common. While its downside is the impossibility of collecting data or changing the colour settings, its benefit as compared to radiometric video is the ease of showing the videos on any laptop or uploading them on a website, making it practical for communication. Some types of software, such as FLIR Thermal Studio Pro and FLIR Research Studio, also allow for converting radiometric video files to non-radiometric ones.

9.4.4 Preventing Damage to the Thermal Camera

The coating on the lens of most thermal camera models is easily damaged by touch. In addition to ensuring that the researchers themselves do not accidentally touch the lens, the thermal recording sessions will have to be planned so that the pigs will not be able to touch the lens or interact with the camera in any way. Thus, the location of the thermal camera should be planned so that it is physically impossible for the pigs to reach the camera (or the tripod, if it is mounted on one), and the lens cap should be kept on whenever the camera is not recording.

In outdoor settings, or when carrying the camera from one place to another outdoors, it is also necessary to ensure that the thermal camera is not pointed towards the sun, as this will reduce its accuracy in the ordinary temperature range to be measured, until the camera is calibrated for the next time. This can be prevented by keeping the lens cap on.

In some environments, there will be an excessive amount of dust in the air, or the risk of water getting splashed on the camera lens, or other risks of some material getting in contact with the lens. Normal glass (such as imaging behind a window) cannot be used to protect the lens, because the wavelengths of infrared radiation that thermal cameras can measure also happen to be wavelengths that do not pass through glass. There are some materials, though, that do let through most of the radiation at these wavelengths. Manufacturers of thermal cameras sell so-called thermal windows, which are made of glass-like materials, such as germanium, and can be placed in front of the thermal camera lens for protection. A lower-cost option, although less standardised and less transparent, is to use a thin sheet of plastic, such as cling film, that is usually used to cover food. Such plastic can be used in cases where the exact numerical values are not important, e.g. in monitoring for substantial visibly discernible temperature changes caused by inflammation, or when the user has sufficient skills in thermal physics to be able to calculate the compensation needed for that specific type and thickness of plastic.

9.4.5 Selecting the Right Type and Capacity of SD Cards or External Hard Drive

For SD cards, Class 10 cards are recommended (they are recognisable by the number 10 inside a C symbol on the SD card), especially when recording radiometric video. For recording single images, it is also possible to use lower-class cards, though these may cause slightly slower functioning of the thermal camera. For streaming radiometric video to be recorded on an external storage device via a laptop, any external hard drive will work. Using a fast USB port on the laptop (an SS port instead of an ordinary port) is recommended if available. Depending on the amount of available storage space on the laptop and on the file sizes of the thermal videos, in some cases, it is sufficient to stream them to the laptop without using an external storage device.

Calculating the storage speed needed on an SD card, laptop or hard drive will depend on the resolution of the thermal camera, the model of the thermal camera (some older models are less efficient in file compression than newer ones), the frame rate of thermal video and the duration of the thermal videos. The easiest way to determine the necessary quantity of storage space is to record a few thermal images or video files with the camera model to be used, with the video frame rate settings at the intended rate, to determine the file size per image or per minute of thermal video.

9.4.6 Charging the Batteries

Thermal cameras operate on rechargeable batteries. Some of the lowest-cost cameras have an inbuilt battery that is charged by plugging in the charging cable to the camera itself. Most thermal cameras offer two options: charging the battery inside the camera as above or removing the battery and charging it in a separate battery charger. Such thermal cameras are normally sold with two or sometimes three batteries, and extra batteries can also be bought separately. To ensure there is always a spare battery when needed, it is recommended to have at least two batteries, so that one can be charged while the other is in use. If a thermal camera is unused for a prolonged time, such as months, it is recommended that the battery be removed to prevent damage to the thermal camera from the rare occurrence of battery leakage or corrosion.

9.4.7 Disabling the Camera Laser

Most thermal cameras have an inbuilt laser, which is used in some industrial settings as an accurate way to measure distance. If the laser beam is accidentally directed at a pig's eye, this can cause eye damage. Therefore, it is strongly recommended that the laser be turned off before using thermal cameras with any animals. If the camera has been used by someone else, the next user should also check that the laser is still off. In most thermal cameras, laser options will be found in the Settings menu (Sect. 9.4.11).

9.4.8   Habituating the Pigs

While a thermal camera records temperature data at a distance, this does not automatically mean it will not disturb or stress (ATOL_0002301) the pigs. The sounds and smells of the camera will be new to the pigs, which can cause anxiety, especially if the pigs are kept in an environment with limited sensory stimuli and/or have had negative experiences with devices such as veterinary equipment. Conversely, pigs may find the thermal camera so interesting that the resulting emotional arousal will affect the results of studies on the autonomic nervous system. For both reasons, it is recommended to habituate the pigs to the presence of a thermal camera and the researcher(s) prior to the study. Habituation will involve first spending time at a distance from the pigs with the thermal camera turned off, then gradually moving closer to the pigs with the thermal camera running (as its sounds and possibly also smells will be different when it is turned on vs. off) but pointing the camera away from the pigs and finally gradually turning the lens towards the pigs. As in any habituation process, the behaviour of the pigs provides the measure for deciding when to proceed to the next step in the habituation process. When the pigs are not reacting behaviourally to the camera, i.e. not standing immobile or erecting their ears as a response to the researcher's presence or movements or the camera noises, it is possible to proceed in approaching. Depending on the anxiety level of the pigs and their prior experiences of similar devices, the habituation process may take from some minutes to several days. It is important not to rush the process, and not to proceed according to some pre-set timetable, because approaching anxious pigs too rapidly will slow down the habituation process, and in worst cases, can even increase the fear of the animals towards the thermal camera. The same habituation process is required for any other equipment, such as camera tripods. It is often easier and less stressful for the pigs if the tripods are brought in and out of the room in the extended state, to avoid causing potentially startling noises near pigs.

9.4.9   Identifying Individuals

In most research and monitoring uses, it is necessary to know which individual is in which thermal recordings. Because thermal cameras do not detect the wavelengths of visible light, any markings made on the pigs' skin with a marking pen or spray will be invisible to a thermal camera. The same is true of the colours of pigs belonging to patterned or fully pigmented breeds.

If the research question or monitoring task is of a type where it suffices to have single thermal images, there are several options for storing information on individual identities. Most thermal camera models come with an inbuilt low-resolution normal camera, which can take an ordinary photograph right before or after the thermal one. This can be activated in the camera menu; in most models, it will be found under Camera settings, submenu Save options and storage (Sect. 9.4.11). Each pressing of the record button will result in one thermal image and one ordinary photograph. Another option, available in a smaller selection of thermal camera models, is typing notes or dictating voice annotations in connection to the thermal image right after taking it.

When recording radiometric video instead of single images, and also in several cases with single images, a separate log of individual identities will be needed. There are several options, such as dictating on a phone or having an assistant write down information on a notepad. One option that requires no extra supplies or staff is to develop a system in which the experimenter's fingers appear briefly at the beginning of the thermal video, signifying the number of the individual to appear on that video. For example, one finger in the upper left-hand corner of the video can represent number 1; two fingers, number 2; three fingers, number 3; one finger in the lower left-hand corner can represent number 4 and so on. Double-digit numbers can be produced by first inserting the fingers to represent tens, for example, two fingers in the lower left-hand corner to represent 50 and then inserting the fingers for single digits, for example, one finger in the upper right-hand corner to represent 7, to signify pig number 57. Such a system will take a while to memorise and practice, but if there are large numbers of animals to record radiometric video on, this is the quickest and easiest way to assign individual identities to thermal videos without the need for an assistant and without the need to talk in the presence of the pigs, in case the setting requires silence.

If pigs are marked for the purpose of the experiment or monitoring, best practices for doing so should be followed. This will help prevent toxic effects, minimise the stress of animals and avoid confounding the results of the study.

9.4.10 Practicing Focusing

One of the prerequisites of accurate data collection from thermal images is the sharp focus of the thermal image. If the image is not in focus, the temperature data in adjacent pixels will "bleed" together, affecting the values in both. Some of the lowest-cost thermal cameras produce images that are in equal focus regardless of distance, but in most thermal cameras, the user will have to focus the image before starting to record. This is usually done by manually moving the focusing ring on the camera lens. Some thermal cameras also have an auto-focus option, which can be used with non-moving or slowly moving animals, but usually reacts too slowly to be practical with fast-moving animals. Rapid manual focusing of a thermal camera is a motor skill that requires practice. Once the animals have been habituated to the presence of the thermal camera and researcher, it is recommended to practice focusing on animals to ensure the skill is at a sufficient level to obtain sharp images before starting data collection.

9.4.11 Checking the Camera Settings

Several of the settings on the thermal camera menu will remain the same throughout a research project or monitoring process, so they will need to be inserted only once. If there is more than one person using the thermal camera, it is also useful to check these settings when the camera returns from another user, if there is a possibility that the other user may have changed the settings.

The settings that will usually remain the same throughout the recordings are listed below. The naming of the settings follows that in the FLIR T-series thermal cameras. Other manufacturers and models may have slightly different naming systems for the corresponding settings.

- Camera settings (the cogwheel icon).
  - Connections: Select your Wi-Fi network if you are streaming radiometric video from the camera to a laptop over Wi-Fi or Bluetooth. This is not needed if your camera is capable of storing radiometric video on an SD card, or if you are taking single thermal images only.
  - Camera temperature range: For maximum accuracy, select the narrowest range that includes mammalian body temperatures, e.g. -20 to 120 degrees.
  - Preview image before saving: In most cases, it is better to keep this off, as the preview will take time before the camera is ready to take the next image.
  - Image resolution: the options here usually are Normal and some form of artificially increased sharpness, such as UltraMax in FLIR or SuperIR in HIKMICRO. This will only affect the visual appearance, not the numerical data. The effect is quite small, but especially if there is a need to communicate with thermal images to external parties, using the enhanced mode will make small details slightly easier to see.
  - Video compression: This setting will determine whether the thermal camera will produce radiometric video (in which each video frame is a radiometric thermal image, i.e. each pixel in each image will contain temperature data as a number) or non-radiometric video (in which the product is a video with a visible heat map but no numerical data). For radiometric video, select Compression to.csq,.seq or another radiometric video file format. For non-radiometric video, select the ordinary video file format available: in most thermal cameras, this will be.mp4,.mov or.avi.
  - Video frame rate: For most uses with pigs, it is recommended to select the fastest rate available, which often is 30 fps.
  - Device settings: Language, time and units can be set according to your preferences. These will determine, e.g. whether the temperature is expressed as Celsius or Fahrenheit, whether your input for the estimate for distance will be interpreted as metres or feet, etc. It is recommended to also periodically check the time and date seen in this menu, and to correct if needed. The camera has an inbuilt clock, but it will gradually drift from true time, and any periods without battery will stop the internal time from running. If the settings for naming image files are set to include the time and date in the file name, these

will be incorrect if the time and date settings in the camera are incorrect. Lamp and laser: This must always be disabled if the camera can be used around live animals, to avoid accidental eye damage.

- Save options and storage: The "Photo as a separate image" option needs to be turned on if a normal photograph is needed to accompany every single thermal image. If photos are not needed, it is useful to turn this option off, because taking a normal photo creates a delay of a second or so before the camera is ready to take the next thermal image. The "Digital camera" option needs to be turned on if normal photographs are needed. The "Measure distance" option will have to be turned off if the method to measure distance uses the built-in laser.

- Focus: Some models offer the option of automatic focusing (autofocus), based on either laser or contrast. Laser must not be used in the presence of live animals, for the above-mentioned reasons, but contrast-based autofocus is safe around animals. Whether or not to use autofocus depends on how fast the animals are moving (faster animals are more difficult for the autofocus) and how much practice the user has had in rapid manual focusing. Some models also offer the option of continuous autofocus. This can be practical if the animal is moving but not rapidly and stays in approximately the same place in the camera view throughout the recording.

- Display settings: Screen rotation will allow for selecting a landscape or portrait orientation for the images. Image overlay information will allow for including selected parameters, (e.g. date and time, air temperature or compass direction) to be visible on each image or on each video frame. Screen brightness will need to be increased if the environment is so brightly lit that it is difficult to see the thermal image on the camera display. Increasing screen brightness will also increase energy consumption and empty the battery faster, so in low light it is possible to extend battery life by reducing screen brightness.

• Recording mode (the camera icon): Select either Single shot (for single radiometric thermal images), Video (for either radiometric or non-radiometric video) or Time lapse (if recording single images at a frequency of full seconds, minutes or hours). If selecting Time lapse, open the subsequent menu to select the intervals and duration of recording.

• Measurement parameters (the icon with three horizontal lines):

- IR window compensation (the icon with two concentric circles): This will normally be set as Off. In the rare case of using an IR window (a specific sheet of a glass-like substance that is mostly transparent to infrared radiation), the input for this parameter represents the loss of infrared radiation in this specific material; this information will have come with the window when purchasing it. If using the lower-cost option of a cling film or other type of thin plastic sheet, this value will have to be determined via experimentation and calculation.

- Emissivity (the ε icon): The parameter expressing how much of the infrared radiation arriving from the surface will be emitted by the surface itself, as opposed to having been reflected from the environment. This is a characteristic of each different type of surface material. The emissivity of human skin is 0.98. The emissivity of different types of animal fur coats varies according to how glossy or coarse their surface is. As there is no published data so far on the actual emissivity of most types of animal fur, it is currently customary to use 0.95 as an approximate emissivity for fur-covered body parts. Therefore, pigs with no hair or very sparse hair will require an emissivity value of 0.98. Pigs of hair-covered breeds, such as the Mangalitsa or Kunekune, will require an emissivity value of 0.95 if the region of interest is in the hair-covered region and 0.98 if the area of interest is in a hairless region, such as the snout disc.

- Image mode (the icon with two squares): In nearly all cases, the option to select is Thermal. The option Digital camera will give a low-resolution normal photograph only, without a thermal image (if the user wants both, that option will be found in Device settings, under Save options and storage). The options Thermal MX and Picture are combinations of a thermal image and a normal photograph, used to find the object of interest in some industrial and electronic settings where the object is of the same temperature as its environment, but these options are nearly always unnecessary when imaging animals with a surface temperature higher than that of their environment.
- Measurement (the crosshairs icon): This setting will determine whether one of the measurement tool options (graphic tools to select specific pixels in the image) is visible during imaging. In most cases, it is easier not to have them visible (by selecting the option "No measurements") to avoid distracting the user when focusing on the animal. All of these tools and many others will be available afterwards, when opening the image with analysis software.
- Colour (the palette icon): This will determine which selection of colours is available to represent different temperatures. Such selections of colours are called palettes, such as white hot (a greyscale image in which warmer temperatures are represented with lighter shades), arctic, lava, iron (all of which are similar to white hot, but with a few added colours to enhance the difference) or rainbow high contrast (a palette of ten colours). It is recommended to select the palette so that, looking at the camera display, it is easy to find the individual animals, see the region of interest and see whether the image is in focus. (These can be fine-tuned further by selecting which parts of the palette will represent which ranges of temperature; depending on the camera model, this is usually done either with a joystick or on the right-hand edge of the touchscreen.) If the images or videos to be recorded are radiometric, the colour palette can always be changed afterwards when the images or video are opened on a laptop with analysis software, so it is not necessary at this stage to think about optimal colour settings for external communication. However, when recording non-radiometric video, it is essential to select the palette and the other colour settings before imaging, as they cannot be changed afterwards.

9.5 Description of the Procedure

9.5.1 Starting a Recording Session

- It is recommended that the thermal camera be turned on at least 15 min before the beginning of the recording session. This will allow the internal temperature of the camera to stabilise, as its internal processor also generates heat.
- During imaging, the camera will auto-calibrate at times. This causes a cracking sound and a momentary (less than a second) freeze of thermal video. This is normal and not to be worried about.

9.5.2 Camera Settings for Each Recording Session

The settings that will have to be changed one or more times for each recording session are as follows. The naming of the settings follows that in the FLIR T-series thermal cameras. Other manufacturers and models may have slightly different naming systems for the corresponding settings.

- Measurement parameters (the icon with three horizontal lines):
  - Distance (the icon with arrows up and down): The distance between the pig and the camera will have to be known or estimated before starting the thermal recording. If the distance changes substantially during imaging, the recording needs to be stopped, the value for distance changed and the recording started again. What constitutes substantial change will depend on whether the data points that will be used in analysis will be within-image differences in temperature (which is recommended to eliminate the error caused by between-image random fluctuations in measurements) or whether the data points for analysis will be raw data, i.e. absolute temperature values as such being compared between different images. The latter is much more sensitive to errors caused by a difference between the settings in the camera and the actual distance. There is little published research so far on how large a difference will introduce a statistically significant error in either case, but as a rough estimate, it is advisable to avoid deviating from the real distance by more than 50% if using within-image comparisons as data points and more than 10% if using raw data as data points.
  - Ambient temperature (the icon with a thermometer bulb and a cloud): The reading of a thermometer, or the temperature reading of a weather meter. It is recommended to measure the ambient temperature at approximately the same height where the pigs are, in case there may be differences in room temperature at different heights.
  - Reflected temperature (the icon with a thermometer bulb and a bouncing arrow): The same reading of the thermometer as above. This parameter will only differ from air temperature if measuring temperatures of reflective

surfaces, e.g. sheets of metal. With high-emissivity surfaces such as animal skin or hair, the effect of reflection is negligible, allowing for the use of ambient temperature as this value as well.

– Air humidity (the icon with raindrops): The reading of a hygrometer, or the air humidity reading of a weather meter. If using a weather meter, the air humidity reading is not to be confused with the other readings that have degrees Celsius or Fahrenheit as the unit, such as dew point temperature. When measuring air humidity, the hygrometer will have to be kept at an arm's length from the measurer's face, because moisture in exhaled air can otherwise affect the reading.

9.5.3 Procedures During the Recording Session

- The thermal camera should be kept as still as possible during imaging. Using a tripod is recommended if feasible. This will minimise the risk that the movement of the camera itself will cause blurriness and therefore inaccuracy of data in the images.
- If it is necessary to turn the thermal camera to follow an animal that has walked out of the field of view of the camera, it is recommended to plan so that the camera moves for as short a time as possible. Reliable data collection is not possible from the thermal videos or images that have been taken during camera movement, for the above-mentioned reason.
- The sharpness (focus) of the focal animal in the camera display will have to be observed throughout the recording session to ensure sharp focus at all times, focusing the camera when needed.
- If recording radiometric video, it is recommended to keep the duration of each video at a maximum of approximately 5 min, as some laptops and some versions of analysis software can be slow to open thermal videos with a longer duration. If a recording longer than 5 min is needed, it is recommended to stop and restart the recording to create separate files.
- To prevent the battery from running out in the middle of imaging, it is recommended to keep an eye on the battery icon in the corner of the camera display and to change the battery when low.
- To prevent the SD card from getting full in the middle of imaging, it is recommended to keep an eye on the SD card icon in the corner of the camera display and to change the card before reaching the capacity limit.
- To prevent damage to the camera, it is necessary to follow the relevant precautions listed in the preparations phase (Sect. 9.4.4).
- Once the recording session is finished, it is recommended to make a backup copy of the recorded files and to store it in a different location from that of the original files.

9.6   Compliance with the 3R Principles

Of the three R's—replacement, reduction and refinement—infrared thermography primarily contributes to refinement, as it is a non-invasive, non-contact technology. For example, if a research question requires the detection of an increase in heart rate (ATOL_0000796) but does not require specific numerical heart rate data, the underlying activation of the autonomic nervous system can be detected by measuring peripheral temperature in the inner corners of the eyes or other peripheral regions that vasoconstrict simultaneously with an increase in heart rate. While heart rate monitors themselves are often non-invasive as well, infrared thermography is also non-contact, which removes the need to habituate animals to heart rate monitors. This is also important as recording from heart rate monitors often requires separating animals from social groups to ensure that companions do not bite or damage the heart rate recording equipment. A non-contact means of measuring physiological processes allows animals to remain in social groups during the measurement.

With more advances in the development of methods for using infrared thermography to measure physiological processes indicative of emotional states and health, the technology may also be able to provide more advances in refinement. At present, however, using infrared thermography in animal welfare science is at an early stage of development and is not yet ready to fully replace other research methods.

9.7   Conclusions

The advantages of IRT technology include its non-invasive, non-contact nature. If the imaging sessions and the preceding habituation are properly planned and carried out, the measurements can be made without disturbing the animals—and, in some cases, even without their noticing. Another advantage is the ability to detect and monitor many physiological processes, provided they involve measurable changes in the animal's surface temperature. The pig is an especially suitable species for such studies, as most pigs have very sparse hair, allowing the skin temperature to be visible to the thermal camera across nearly all body parts.

The main disadvantage of this technology is the multifactorial nature of surface-temperature changes in thermal images. Factors that can influence the surface temperature distribution of an animal include its emotional state; current and recent physical exercise; metabolic processes; diurnal rhythm; oestrus cycle; health status, such as inflammation responses and fever; ambient temperature; air currents; nearby heat sources; and so on. Disentangling the temperature effects of the physiological process of interest from those caused by other factors requires expertise and, in many cases, a standardised environment. Another disadvantage, especially for researchers new to the technology, is the seemingly simple and attractive appearance of thermal images. This has, in some cases, led to failed research projects, e.g. when the researchers have not had enough understanding of the multitude of factors that can influence surface temperature, or of the thermal dynamics of the processes they are studying. It is therefore recommended that any research project in thermal imaging involves at least one researcher or advisor with sufficient prior experience in thermal imaging, an in-depth understanding of thermal physics and the workings of a thermal camera, as well as all the physiological processes that can alter surface temperature distribution in the species in question in the circumstances of the study. It is also recommended that the development of new methods to use thermography to measure animal welfare and health be continued, because the field as a whole holds substantial promise for the future development of new and useful methods for research and monitoring, if done skillfully.

References

Franchi GA, Moscovice LR, Telkänranta H et al (2024) Variations in salivary oxytocin and eye caruncle temperature indicate response to environmental enrichment material in fattening pigs. Appl Anim Behav Sci 275:106291. https://doi.org/10.1016/j.applanim.2024.106291

Ghezzi MD, Ceriani MC, Domínguez-Oliva A et al (2024) Use of infrared thermography and heart rate variability to evaluate autonomic activity in domestic animals. Animals 14:1366. https://doi.org/10.3390/ani14091366

Kammersgaard TS, Malmkvist J, Pedersen LJ (2013) Infrared thermography—a non-invasive tool to evaluate thermal status of neonatal pigs based on surface temperature. Animal 7:2026–2034. https://doi.org/10.1017/S1751731113001778

Minkina W, Dudzik S (2009) Infrared thermography: errors and uncertainties. Wiley

Rekant SI, Lyons MA, Pacheco JM et al (2016) Veterinary applications of infrared thermography. Am J Vet Res 77:98–107. https://doi.org/10.2460/ajvr.77.1.98

Stewart M, Verkerk GA, Stafford KJ et al (2010) Noninvasive assessment of autonomic activity for evaluation of pain in calves, using surgical castration as a model. J Dairy Sci 93(8):3602–3609. https://doi.org/10.3168/jds.2010-3114

Image Analysis for the Assessment of Pig Body Composition

Maria Font-i-Furnols, Tamás Donkó, Patrick Schlegel, Ludovic Brossard, Marina Gispert, Albert Brun, Catherine Ollagnier, and Claudia Kasper

Abstract

The use of less invasive technologies to determine the body composition of pigs during growth is important for nutritional, breeding and management purposes. Nowadays, technologies based on ultrasound, X-rays and nuclear magnetic resonance are available and can be used either directly on the farm for selection or strategy decisions, or in a laboratory for research purposes. This chapter describes the most common non-invasive technologies used to characterise body composition, addressing the aspects to consider for both the preparation of the animals and the acquisition and analysis of measurements. It also considers the differences between techniques and the information that can be obtained from each technique.

Keywords

Non-invasive · Ultrasounds · Computed tomography · Dual X-ray absorptiometry · Nuclear magnetic resonance imaging

M. Font-i-Furnols (✉) · M. Gispert · A. Brun
IRTA-Food Quality and Technology, Monells, Catalonia, Spain
e-mail: maria.font@irta.cat

T. Donkó
Medicopus, Kaposvar, Hungary

P. Schlegel · C. Ollagnier
Swine Research Group, Agroscope, Posieux, Switzerland

L. Brossard
PEGASE, INRAE, Institut Agro, Saint Gilles, France

C. Kasper
Animal GenoPhenomics Group, Agroscope, Posieux, Switzerland

© The Author(s) 2026

129

L. R. Moscovice et al. (eds.), *Standard Operating Procedures for Better Pig Research (SOPig)*, https://doi.org/10.1007/978-3-032-24899-2_10

10.1   Introduction

The body composition assessment of live pigs provides information regarding the development of tissues or the accretion of nutrients during growth (e.g. fattening pig) or contents between changing physiological phases (e.g. sow) (Gonzalo et al. 2018; Heurtault et al. 2024a; Floradin et al. 2024; Font-i-Furnols et al. 2020; Zomeño et al. 2016; Carabús et al. 2015; Kremer et al. 2013). This information represents the basis for developing the growth curve of tissues that are necessary to define nutrient requirements for a desired tissue composition (lean, fat or bone tissues) at a desired time point. Body composition is also a parameter that can be considered when differentiating, for example, breeds, new genetic lines, techniques of castration or diets and can be used to determine feed efficiency (ATOL_0001580). It can be determined destructively—that is, by slaughtering the pigs—or non-destructively and non-invasively by using technologies based on ultrasound, X-rays or magnetic resonance. These technologies can estimate either the chemical or physical composition of the tissue (after a calibration process) or its anatomical distribution.

Ultrasound is based on acoustic waves propagated through materials as perturbations in their physical structure that return to the transducer and are converted into images. In the A-mode (amplitude modulation) ultrasound, a single transducer emits a pulse in the body, and the returning echoes are displayed as a function of depth. In B-mode (brightness modulation) ultrasound, i.e. 2D mode, a linear array of transducers simultaneously scans the body surface, returning a 2D image to the screen. Both have been applied to swine research and production for the last 20–25 years. B-mode and real-time ultrasound have become the preferred methods (Carabús et al. 2016).

Computed tomography (CT) is based on X-rays, which are attenuated to varying degrees as they pass through the carcass, depending on the density of the tissues. CT provides a discrete 3D image of the body consisting of multiple voxels (3D pixels). The dimension of the voxel is defined by the settings of the CT scanner, for example, the matrix, the field of view and the image thickness. Each voxel is associated with a Hounsfield value (HU) that corresponds to the average attenuation in this volume according to the Hounsfield scale (air $= -1000$ HU and water $= 0$ HU). In live pigs, fat has attenuation values between -1 and $- 149$ HU, lean (or muscle) has values between 0 and $+ 140$ HU and bones have more than $+140$ HU approximately. Heart, liver and spleen have HU values similar to that of muscle, while lungs have more negative values than fat. CT produces an initial radiographic image (scout or surview) used to select the region to be scanned—either the whole animal or just a specific region.

Dual X-ray absorptiometry (DXA) is, as CT, based on X-ray emissions through the body, which are attenuated depending on the density of the body tissues (lean mass, fat and bone) they pass through. DXA scans provide 2D images distinguishing three types of body tissues (lean mass, fat and bone). DXA allows scans of the complete body or of specific regions (e.g. spine). The results of a scan include the weight of the lean tissue, weight of the fat tissue, weight of the bone mineral tissue, surface area of the skeleton and areal bone mineral density. The total lean tissue mass, fat tissue mass and bone mineral mass equals the total mass. Within the DXA scan image, regions of interest (ROI) are defined, and the results are presented for each ROI separately as well as for the total scan region.

The basic principle of nuclear magnetic resonance imaging (MRI) is that atomic nuclei with an odd number of protons and/or neutrons absorb and reemit radio waves when placed in a magnetic field. As the most abundant atom in the body, hydrogen exhibits a magnetic moment. By contrast, elements such as carbon (C) and oxygen (O) lack a magnetic moment due to their even number of protons and neutrons. The presence of a magnetic moment causes hydrogen protons to emit a significantly stronger signal. This characteristic makes the hydrogen nucleus an appealing isotope for imaging purposes. To align the protons that are normally randomly oriented, a homogeneous magnetic field is needed (0.28–3 Tesla for a whole-body scan). The radio wave frequency at which the nuclei resonate is directly proportional to the strength of the magnetic field and is called the Larmor frequency (for protons, 64 MHz at a field strength of 1.5 T). The response signal of the protons varies based on the tissue in which they are embedded because of the bonded atoms affecting the electromagnetic field. This effect can be exploited to differentiate protons in one tissue from those in others. Following the transmission of a radio frequency (RF) pulse, the protons in the tissue release the absorbed energy, generating a small RF signal that can be detected and used for cross-sectional image reconstruction (Baulain 1997).

10.2 Goal and Scope of the Procedure

The goal of this procedure is to describe the most common non-invasive methodologies used to characterise the composition of the body of live pigs, mainly in terms of morphometric measurements, fat and lean content. The methodology describes the steps of prerequisites and preparation, image acquisition and the following image processing.

10.3 Materials and Equipment

10.3.1 Ultrasound, CT, DXA and MRI

For ultrasound, the choice between a mode A or a mode B ultrasound depends on the type of information needed, e.g. whether fat and muscle thickness or area measurements are required. In both cases, a gel is necessary to ensure proper contact between the probe and the pig's skin (ATOL_0005636). Additionally, if the animal has hair, materials for hair removal are required to facilitate good contact with the probe.

The CT scanner is an X-ray device that must be installed in a lead-lined room. This room must be constructed in compliance with legal regulations to ensure worker safety. Additionally, dosimetry badges and protective protocols are safety measures needed to monitor and control radiation exposure to personnel. Moreover, the installation and operation of the CT scanner are subject to official inspections and oversight by authorised personnel, with the frequency of these controls determined by prevailing regulations. It is essential to consult the specific regulations applicable in each country, as legal requirements may vary and evolve over time.

The specifications required of the equipment depend on the intended use. Generally, the lower the number of slices, the faster the scanning process and the lower the radiation dose. The diameter of the gantry, the length of the table and maximum weight allowed, and the scanning coverage should also be considered to ensure it accommodates the diameter and length of the animals to be scanned. Sometimes a foam can help to separate parts of the pig (e.g. the legs) or, if the pig is placed in a tube, to separate the pig from the tube, in order to facilitate the subsequent image analysis. Also, foams can help to immobilise the tube in the CT table.

For the DXA device, although it is an X-ray device, the lead-lined room is not needed since the radiation is lower than with CT. However, it is important to keep the safety distance while scanning to ensure worker safety. Individual badges for irradiation control are recommended, and, like CT, official inspections are also carried out in agreement with the current legislation of each country.

To choose a DXA device, the distance between the table and the receptor, the table size and scanning characteristics should be considered to ensure it accommodates the weight, diameter, height and length of the anaesthetised pigs to be scanned. It is important that the table allows to fit all the animals to ensure they are completely within the scanned region, which is marked by a white line. Sometimes, to have a good quality of the images and avoid the body overlaps with the limbs, it is important to have some foam positioning aids in order to use them to separate the front legs from the body.

During in vivo and post-mortem porcine MRI examinations for body composition analysis, the availability and proper calibration of high-field magnetic resonance imaging equipment are essential, operating at a 1.5 T or 3.0 T magnetic field strength. The scanner should provide whole-body imaging capacity with appropriate core dimensions to accommodate the animal's size, as well as dedicated coils that ensure sufficient signal-to-noise ratio and spatial resolution. All auxiliary devices, such as monitoring systems, anaesthesia machines and infusion pumps, must be fully MRI-compatible to prevent interference with image quality and to comply with magnetic field safety requirements. Foam positioning aids, as used in CT and DXA, can assist in properly positioning the pig.

Safety considerations are of critical importance in both the in vivo and post-mortem settings. Strict adherence to MRI safety standards is required to prevent hazards associated with ferromagnetic materials, radiofrequency heating and gradient-induced peripheral nerve stimulation. All personnel involved in animal handling must be trained in MRI-specific safety protocols, including screening for incompatible equipment and personal items. The scanning environment must ensure controlled access, continuous supervision and emergency preparedness, with resuscitation equipment located outside the magnetic field zone.

Table 10.1 Overview of equipment needed and aspects to consider regarding different image analysis devices and their installations

Equipment needed and aspects to consider	Image analysis method			
	Ultrasound	*CT*	*DXA*	*MRI*
Gel to ensure proper contact of probe	X			
Equipment for hair removal	X			
System to transport the pigs to the scanning table		X	X	X
Full anaesthesia needed		X	X	X
Foam positioning aids		X	X	(X)
Individual radiation badge to control radiation		X	X	
Keep safety distance during scanning		(X)	X	
System to limit the diameter of the pig (e.g. half of a cylindric tube, conforming cover, etc.)		X		X
Operational room with armour during examination		X		
Shielded scanner room (Faraday cage)				X
Scanner room and material with no metallic pieces				X
Distance of the material from the magnetic field with cardiac pacemaker, defibrillators, insulin pump, implants made of metal, hearing instruments, dosage devices for medication, etc.				X

For CT, DXA and MRI, a transport system is needed to move pigs to the table of the scanner, especially if they are heavy. This system could be an overhead rail, a wheeled table or a hoist (Fig. 10.3c in Sect. 10.5). Particular attention must be paid in MRI to ensuring that no magnetisable parts of these animal movement tools come into contact with the equipment.

In CT, the image acquisition occurs within a circular field of view and, similarly, in MRI, the homogeneous region of the magnetic field is also circular. Therefore, using half of a cylindrical tube to position the animal or a conforming cover to wrap the pig is recommended. This helps to prevent the scanning diameter from being too large and ensures the pig fits within the effective imaging area. Adequate data storage and processing units are also necessary for the acquisition and evaluation of volumetric datasets. A summary of the equipment and installations needs is presented in Table 10.1.

10.3.2 Sedation and Anaesthesia Material

Anaesthesia is required to minimise stress (ATOL_0002301), pain (ATOL_0000863) and motion artifacts during the scanning procedure with CT, DXA and MRI. Type of sedation/anaesthesia applied depends on the objective and the characteristics of the scan to be performed such as duration of scanning and region scanned. Anaesthesia protocols should be tailored according to the animal's body weight (ATOL_0000351) and physiological condition, typically involving premedication, induction with injectable agents and maintenance with inhalation anaesthetics.

For anaesthesia based on intramuscular or intravenous injections, the required materials include needles and syringes. A catheter (or IV line) can help administer the injection if the animals move. The size of the syringe depends on the dose to be administered. Additional small materials for disinfecting and cleaning blood (ATOL_0005631) are also necessary.

For volatile anaesthesia, a device for inhalation of anaesthesia is necessary, with all the complementary material such as mask (size according to the size of the pig), tape to fix the mask to snout, isoflurane, oxygen and soda lime granules. This needs to be accomplished with all the safety requirements.

Since body temperature decreases under anaesthesia, thermometers are required to monitor it. Additionally, blankets or heating pads should be available to warm the pigs during recovery if necessary.

10.4 Prerequisites and Preparation

Personnel responsible for managing the devices must receive proper training to know how and where to measure and to ensure reliable information. In some countries, operating X-ray or magnetic resonance equipment requires specific courses and certification. For safety reasons, working with X-ray devices requires knowledge of radiation, and for MRI equipment, training on materials compatible with the technology is needed. Therefore, workers must comply with the legal requirements applicable in each country.

10.4.1 Sedation and Anaesthesia Protocols for Pigs

Except for ultrasound, all other non-invasive imaging technologies require that the animal remains in a defined position during scanning. Thus, the animal needs to be anaesthetised. To ensure the safety and welfare (ATOL_0000765) of the animals, a veterinarian should be consulted to select the appropriate anaesthesia protocol and to train the experimenter on anaesthesia techniques. Moreover, it is advisable to have a veterinarian present during the entire process to take action in case of an emergency. The administration of anaesthesia requires knowledge that includes the monitoring of animals during anaesthesia, the handling of emergencies and the handling of volatile anaesthesia machines. For animal welfare, the thermal comfort of the anaesthetised animal is of particular importance, since anaesthesia downregulates thermal regulation. Animals must also be monitored until they are fully awake.

Anaesthesia type and doses have to be adapted to the time the animal needs to be anaesthetised. For instance, an MRI scan usually takes substantially longer than a CT scan (although it depends on the characteristics of the CT scanner). A cross-sectional scan of a 30-cm range typically takes several minutes. Thus, an MRI examination of an entire pig can take up to 1 hour, depending on the sequence used and the size of the animal.

Table 10.2 Active principles, doses and withdrawal times for drugs used in protocol 1

Active substances	Dose for intramuscular injection	Withdrawal time[a]
Azaperone	2 mg/kg of body weight (BW)	7–18 days for meat
Ketamine	15–20 mg/kg BW	1 day for meat

[a]Withdrawal times may differ depending on country

Three anaesthesia protocols are proposed here as examples, with different durations and withdrawal times (i.e. the time required after administration of a drug until the meat is considered safe for human consumption). The withdrawal time may differ from country to country because regulations are subject to change over time and by country. Therefore, please refer to the product information before using it. In one of the examples (protocol 2, Sect. 10.4.1.2), the animal cannot be used for human consumption according to current EU legislation.

The volatile anaesthesia (protocol 3, Sect. 10.4.1.3) has the advantage that the pig wakes up very fast and can be brought back to its home pen more quickly compared to intramuscular/intravenous anaesthesia. The downside is that it might be more stressful for the pigs, and they often struggle against falling asleep.

10.4.1.1 Protocol 1: Injectable Anaesthesia

This anaesthesia protocol uses products that have a market authorisation (and therefore an official withdrawal time) for swine (Table 10.2). The present protocol may be used as an induction protocol for heavy pigs (>140 kg body weight), but it will not be sufficient on its own: it should therefore be applied in conjunction with volatile anaesthesia (isoflurane). This method induces anaesthesia for 0.5–1.5 h.

To carry out the anaesthesia, two intramuscular injections need to be applied to the pig, with 10 min between each injection. All equipment and the administration of the injection must be carried out under appropriate hygienic conditions. The first injection (azaperone) will calm the animal, and in some cases, the pig may lie down. The second injection (ketamine) will anaesthetise the pig. It is important to monitor the pig's health status (e.g. respiratory rate (ATOL_0001796), body temperature, skin colour (ATOL_0000486) throughout anaesthesia.

Assess the depth of anaesthesia (e.g. muscle tone, especially in the tail and ears) before starting the scan. If the anaesthesia is not sufficiently deep, increase the concentration of ketamine as needed. Approximate time of onset: 15–30 min.

10.4.1.2 Protocol 2: Injectable Anaesthesia with Long Duration

There is no marketing authorisation for the products used in this protocol, so they cannot be used on pigs raised for human consumption. This method induces anaesthesia for approximately 1 h.

To carry out the anaesthesia, one intramuscular injection of the anaesthesia cocktail needs to be applied to the pig (Table 10.3). All equipment and the administration of the injection must be carried out under appropriate hygienic conditions. Monitor the pig's health status (e.g. respiratory rate, body temperature, skin colour) throughout anaesthesia. Assess the depth of anaesthesia (e.g. muscle tone, especially in the tail and ears) before starting the scan. Approximate time of onset: 10 min.

Table 10.3 Active principles and doses for drugs used in protocol 2

Active substances	Dose for intramuscular injection[a]
Tiletamine	1.250–1.875 mg/kg BW
Zolazepam	1.250–1.875 mg/kg BW
Xylazine	1.250–1.875 mg/kg BW
Ketamine	1.250–1.875 mg/kg BW
Butorphanol	0.250–0.375 mg/kg BW

[a]Note that pigs cannot go to slaughter after anaesthesia since there are no official withdrawal times for consumption of the meat

Table 10.4 Active principles, doses and withdrawal times for drugs used in protocol 3

Active substances	Dose	Withdrawal time[a]
Isoflurane	Adapt the dose to the pig's size, oxygen flow and depth of anaesthesia of the pigs	1 day for meat

[a]Withdrawal times may differ depending on country

10.4.1.3 Protocol 3: Volatile Anaesthesia

Volatile anaesthetics, such as isoflurane, can be administered to pigs using a mask placed over their noses. It is a very safe anaesthesia, but its onset is dependent on the size of the pig. The calmer the pig, the easier it is to hold the mask to deliver isoflurane.

Volatile anaesthesia agents such as isoflurane usually have a marketing authorisation (and therefore an official withdrawal time) for swine (Table 10.4). Volatile anaesthesia requires a dedicated inhalation machine that should be checked prior to each use. A veterinarian should be consulted for proper training on how to perform anaesthesia with volatile anaesthetics.

To carry out the anaesthesia, first, the apparatus needs to be set up:

1. Check that isoflurane and oxygen levels are sufficient to ensure sedation throughout the entire scan.
2. Check the saturation of soda lime granules. If they turn purple, they must be replaced.

After the apparatus has been checked, carry out the anaesthesia in the following steps:

1. Open the valves on the oxygen cylinder, then open the valve connecting the oxygen cylinder and anaesthesia device.
2. Restrain the pig securely while positioning the anaesthesia mask on its snout.
3. Once the mask is positioned, turn on the oxygen supply and the isoflurane flow.
   a) Set the oxygen flow rate to 3 L/min.
   b) Set the isoflurane concentration to 5%.
4. Fix the mask on the pig's snout with tape as soon as it is asleep.
5. Once the pig is fully anaesthetised, reduce the isoflurane concentration based on the pig's body weight.
   a) 2–3% for pigs up to 15 kg
   b) 4–5% for heavier pigs.

6. Monitor the pig's health status (e.g. respiratory rate, body temperature, skin colour) throughout anaesthesia.

Assess the depth of anaesthesia (e.g. muscle tone, especially in the tail and ears) before starting the scan. If the anaesthesia is not sufficiently deep, increase the isoflurane concentration as needed. Approximate time of onset: 1 (newborn) to 10 min (140 kg).

10.4.2 Preparation of Pigs Before Examination/Scanning

A checklist with aspects to consider when preparing the animals for ultrasound examination or CT/DXA/MRI scanning is presented in Table 10.5.

10.4.3 Preparation of Ultrasound, CT, DXA and MRI Device

Before using the devices, some preparation is needed in terms of calibration or verification of the correct functioning (Table 10.6).

Table 10.5 Aspects to consider regarding the preparation of the pig before performing image analysis

Pig preparation measure	Ultrasound	CT	DXA	MRI
Fasting time and health conditions of pigs before scanning		X	X	X
Body weight of pig prior to anaesthesia via injection to determine the required amount of sedative		X	X	X
Immobilisation of pigs (before sedation/anaesthesia or examination)[a]	X	X	X	X
Check the effectiveness of anaesthesia before scanning		X	X	X
Application of intravenous or feeding markers to the animal if needed for some specific evaluations		X		X
Removal of hair at the location of examination	(X)			
Application of gel or another substance on the skin surface	X			
Removal of the ear tag or other metallic pieces				X
Single pen prepared to allow calm awakening of the anaesthetised pig after scanning		X	X	X

[a]Training of pigs should be considered for stress-free immobilisation during ultrasound or for application of anaesthetics

Table 10.6 Aspects to consider regarding the devices and the scanning general procedure of the installations/devices

General procedure	Ultrasound	CT	DXA	MRI
Verification of the correct functioning of the device	X	X	X	X
Legal requirements to work (armour etc.)[a]		X	X	X
Warm up of the device before use		X	X	X
Calibration of the device before use		X	X	X

[a]May differ from country to country

Table 10.7 Aspects relative to the position of the pig for the measurement

	Ultrasound	CT	DXA	MRI
Transport of the pigs to the table of the device for scanning		X	X	X
Proper positioning of the pig during examination/scanning (standing up or prone in ultrasound; prone, supine or lateral in CT, DXA and MRI)[a]	X	X	X	X
Correct position of the legs (stretched, under the body, etc.)[b]		X	X	X
The body (part) to be measured must be placed in the measuring area or field so that the mobile parts of the device can move freely		X	X	X

[a]Position has an effect on the movement of the pig due to breathing and, consequently, on the quality of the image. Prone position is the most used. It is important to place the pigs in the same position within a trial to yield comparable results and include foams in specific places to ensure image quality
[b]This may depend on the type of image and the characteristics of the device, such as the maximum length of the measurement

Table 10.8 Aspects to be considered for the scanning or evaluation of the pigs with different devices

Scanning procedure
Definition of the anatomical position/s in which the measurement is to be carried out (one or various specific locations depending on the objective).
Definition of the acquisition parameters (depends on the device capacity and the objectives of the evaluation)

Before scanning or measuring, immobilised or anaesthetised pigs need to be correctly positioned and prepared for measurement. Several procedural aspects that can influence the measurement should be established and defined in the protocol (Table 10.7).

Furthermore, the scanning protocol should be defined based on previous experience, the possibilities of the device, the objective of the evaluation or the literature review carried out. Some aspects to be considered are presented in Table 10.8.

10.5 Description of the Procedure

10.5.1 Ultrasound Measurement

Once the equipment is verified (Table 10.6), the animal is prepared (immobilised and correctly placed with hair removed from the area of measurement) and the gel is applied to the surface of the skin (Table 10.5 and Table 10.7) to allow better contact of the ultrasound probe and facilitate the examination, the measurement can be taken, considering the issues indicated in Table 10.8. For the correct operation of the equipment, the manufacturer's instructions should be followed.

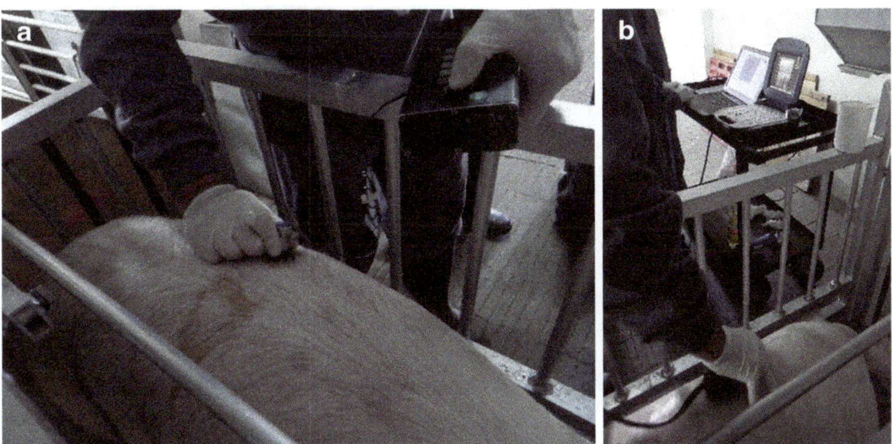

Fig. 10.1 Measurement of a pig with two different types of ultrasound devices: (**a**) A-mode, (**b**) B-mode (Photo credit: Maria Font-i-Furnols, IRTA)

10.5.1.1 Image Acquisition and Image Treatment

The measurement is carried out by placing the probe perpendicular to the place to be evaluated (usually the loin at a specific position defined anatomically). To obtain good ultrasound measurements, good contact between the probe and the tissue is very important (Fig. 10.1).

Measurements should be taken by a trained person and, if possible, in duplicate or triplicate. Depending on the type of ultrasound, two types of information can be obtained:

- Direct measurement of fat thickness (ATOL_0001517) or muscle depth.
- A 2D image where it is possible to measure fat thickness and muscle depth, loin area, etc. (Fig. 10.2).

If calibration equations are available, it is possible to predict the lean meat content from the variables obtained (Lucas et al. 2017). B-mode/medical ultrasound is highly dependent on the type of transducer used (linear, convex, phase array). It can produce both 2D and 3D images and cardiac flow analyses.

10.5.2 CT, DXA and MRI Scanning

Once the prerequisites for the scan (Table 10.6) have been met, the pigs are prepared (Table 10.5) and placed on the scanning table (Table 10.7), and the scan can be carried out in accordance with the established protocol (Table 10.8). For the correct operation of the equipment, the manufacturer's instructions should be followed. For all devices, the measurement is carried out by taking into account all safety rules to reduce or avoid exposure to irradiation.

Fig. 10.2 Ultrasound image of the loin region of a pig

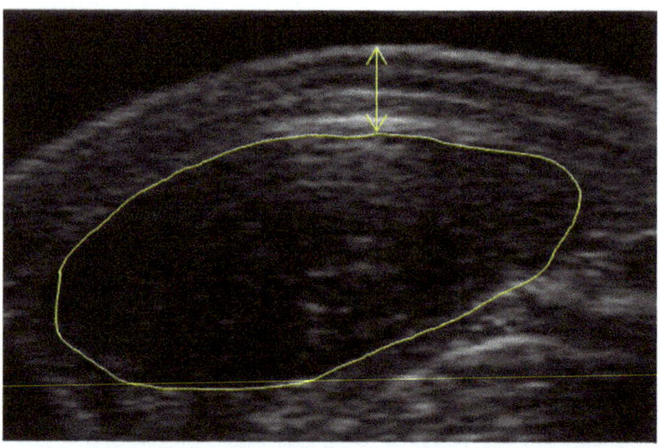

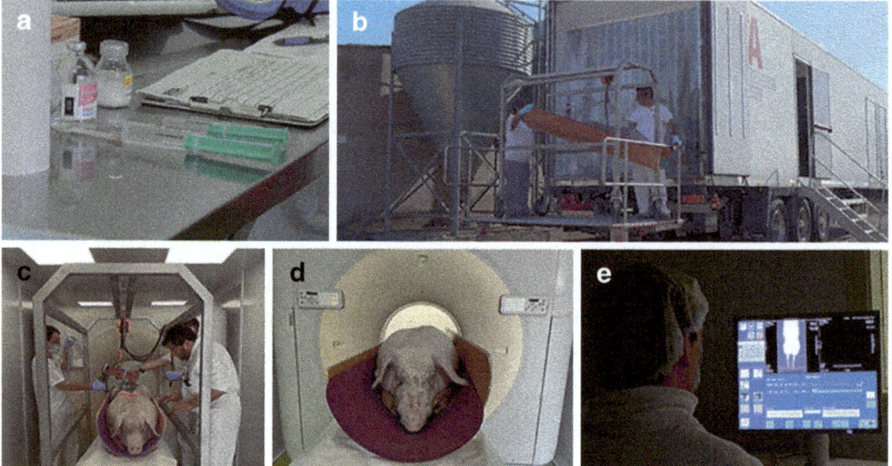

Fig. 10.3 Preparing the pig and scanning with computed tomography: (**a**) anaesthesia, (**b**) transport of the pig to the CT, (**c**) placement of the pig to the CT scan table, (**d**) position of the pig on the scan table for scanning, and (**e**) scanning from the operator room (Photo credit: Maria Font-i-Furnols and Albert Brun, IRTA)

10.5.2.1 Preparation of the Pig: Sedation and Positioning

See Sect. 10.4.1 which includes protocols for anaesthesia, either by intramuscular or intravenous injection (Fig. 10.3) or with volatile anaesthesia (Fig. 10.4). Moreover, Fig. 10.3. shows an example of the process of preparing and positioning the pig in the computed tomography device, which can also be applicable to DXA and MRI (using material suitable for this device).

10.5.2.2 CT Image Acquisition and Image Treatment

Image acquisition is related to the scanning protocol, such as intensity, voltage, thickness, matrix size, pitch, overlapping, collimation, field of view and by the

Fig. 10.4 Standard position of a 100 kg BW pig for complete body acquisition by DXA. Sedation by inhalation of isoflurane (Photo credit: Patrick Schlegel, Agroscope)

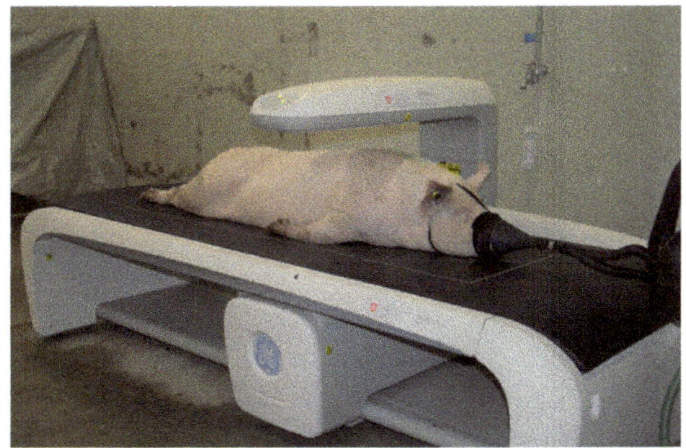

reconstruction algorithm used. The output from the CT scanner is a stack of images, typically in DICOM format (https://www.dicomstandard.org). Each one is, in fact, a matrix of numbers. If a pig of 120 kg is scanned longitudinally (cranial to caudal) and images are taken every third mm, the output would be around 560 images. The acquisition time is very variable and depends on the characteristics of the CT and the acquisition protocol, e.g. it can be less than one minute or more than half an hour.

CT images are calibrated and expressed in Hounsfield units (HU), providing quantitative values that directly correspond to tissue density. Each pixel represents an absolute attenuation coefficient relative to water, allowing for standardised and comparable measurements across scanners.

From the images, it is possible to take linear measurements or calculate areas at different specific anatomical positions (e.g. in Fig. 10.5 and Table 10.9). It is also possible to calculate volumetric parameters by combining multiple images. In this case, the volume associated with each Hounsfield value can be obtained. This allows a classification of each voxel into fat, lean and bone by using threshold segmentation and, hence, quantification of the different tissues (Fig. 10.6). If needed, the viscera, organs and testes (in the case of entire or immunocastrated males) can be removed from the calculations using appropriate software. This would depend on the objective of the study: If the objective is to obtain the characteristics of the carcass from live pigs, it is probably better to remove the viscera, organs and all parts that do not belong to the carcass. If the objective is to obtain the characteristics of the whole body (e.g. protein content), there is no need to remove these items. Moreover, the volume associated with each HU can allow predicting the dissected lean, fat or bone contents, lean %, fat %, chemical protein, fat, ashes or moisture by applying previously obtained calibration equations. It is also possible to determine the proportion of low, medium or high-density bones by considering volumes associated with a different range of Hounsfield values. The type of data required and the way to extract them will depend on the objective of the project.

The presentation of the results is adapted to the type of information obtained from the CT images. A database with the information is obtained to be used for statistical calculations.

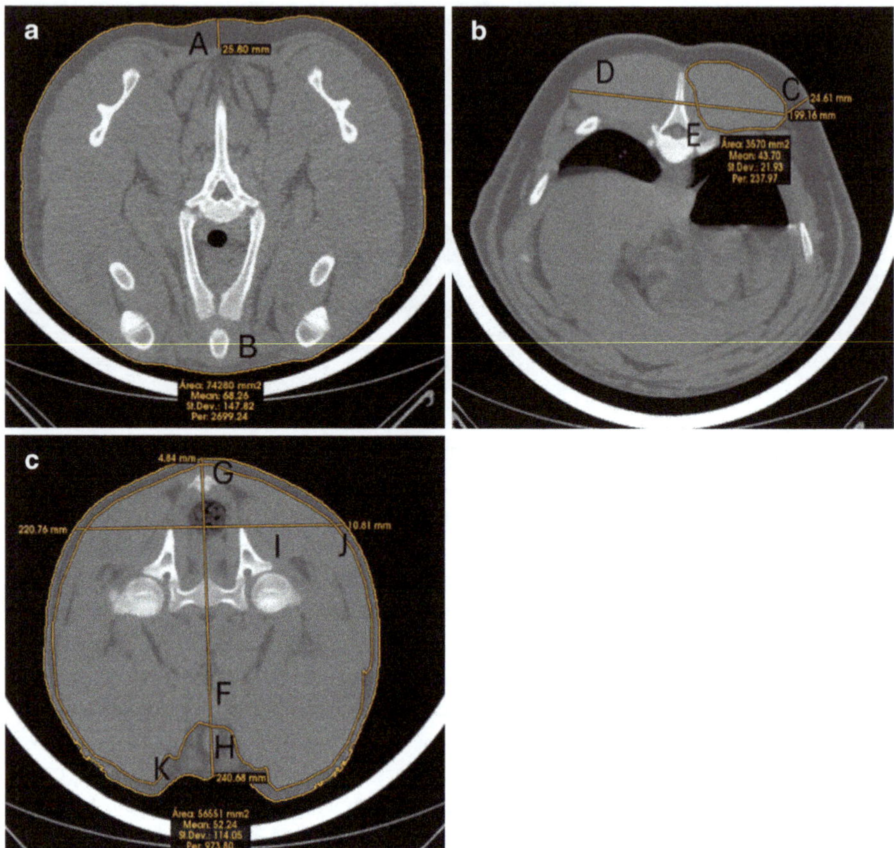

Fig. 10.5 Example of possible anatomical measures obtained from the tomogram of the shoulder (**a**), loin (**b**) and ham (**c**) (source: adapted from Carabús et al. 2015)

10.5.2.3 DXA Image Acquisition and Image Treatment

Depending on the DXA device, several acquisition modes for the complete body are available. As an example, GE Healthcare iDXA has the following modes:

- Total body—thick: mean height of >25 cm.
- Total body—standard: mean height of 16–25 cm.
- Total body—thin: mean height of <16 cm.
- Small animal body—large: > 20 kg.
- Small animal body—medium: 2–20 kg.
- Small animal body—small: < 2 kg.

At Agroscope, all pigs are scanned using total body (thick mode), defined as the standard mode for live animal scans to ensure consistent comparability of body composition across pigs of varying sizes. Alternatively, total body (standard) or small animal body (large mode) settings may be used for pigs with lower body weight.

Table 10.9 Anatomical location of the measurements taken from each tomogram (source: adapted from Carabús et al. 2015)

Tomogram	Measurements
Shoulder (Fig. 10.5a)	Subcutaneous fat thickness perpendicular to the skin measured in the middle of the vertebral column (A)
	Area and perimeter of the whole shoulder image (B)
Loin (Fig. 10.5b)	Lateral fat thickness of right loin eye measured perpendicular to the skin at the most lateral part of the loin (C)
	Maximum length of the 2 loins (D)
	Loin eye area and perimeter (E)
Ham (Fig. 10.5c)	Maximum vertical height of the ham (F)
	Subcutaneous fat thickness measured perpendicular to the skin at the top of the ham (G)
	Area of the subcutaneous fat of the hams (H)
	Width of the ham measured above the bones (I)
	Lateral fat thickness (mm) at the previous level (J)
	Area and perimeter of the ham (K)

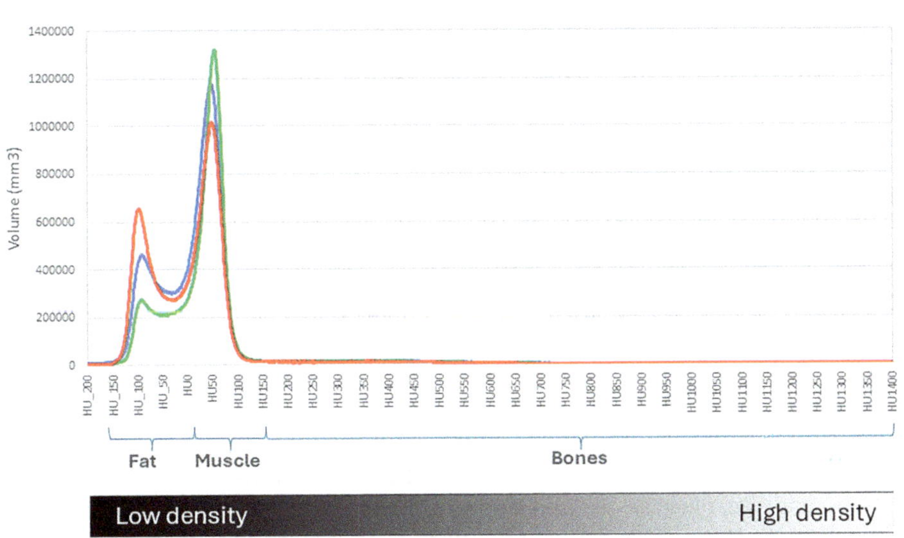

Fig. 10.6 Volume associated with each Hounsfield value of three pigs: a fat pig (red), a medium fat pig (blue) and a lean pig (green)

For repeated measurements over growth, a choice needs to be made: (1) adapt in acquisition modes according to the growth stage and accept the differences in calculation procedures of tissue masses between acquisition modes, or (2) select one single acquisition mode suitable to the highest weight range to be used over the

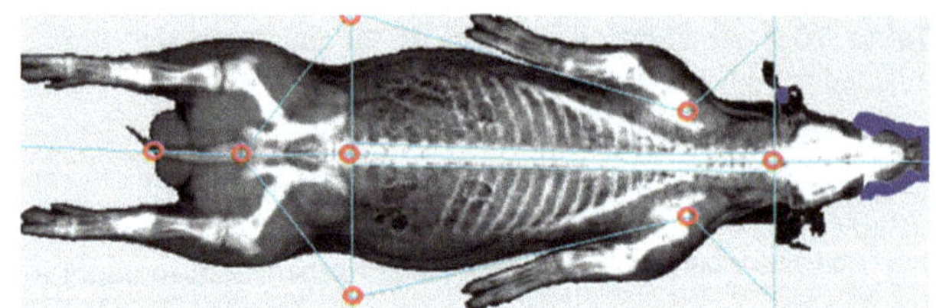

Fig. 10.7 Scan image after positioning the ROI and removing artefacts (Kasper et al. 2021)

complete period and accept to not strictly follow the recommendations for the DXA device.

The acquisition time depends on the scanning mode chosen and the size (length) of the pig. For pigs that are significantly shorter than the DXA table, the scan can be aborted to save time, to avoid scanning the empty part of the table until it reaches the end.

The image treatment can be performed at any time after the acquisition and consists of the following (Fig. 10.7):

1. Adapting the placement of the regions of interest (ROI). As the equations to determine lean tissue, fat tissue and bone mineral masses from the X-ray signals depend on the ROI, the correct placement of the ROI on the body is relevant. This means that the ROI should be placed in a way that is closest to the regions defined for humans. For instance, the ROI lines for "Spine" should be placed to the left and right of the vertebral column, or the ROI for "Left arm" should contain the pig's left front leg.
2. Checking the correct attribution to body tissues. Attribute non-animal parts (ear tag, inhalation mask, inhalation tube…) to "artefact".

Depending on the device, data can be exported as a data file, which is an advantage in research, as the interest is not focused on one single "patient". With GE Healthcare iDXA, the exported results are arranged as shown in an example of one complete body acquisition in Fig. 10.8. A review of the DXA methodology can be found at Heurtault et al. (2024b).

10.5.2.4 MRI Image Acquisition and Image Treatment

Prior to imaging, the anaesthetised animal is placed on the examination table and an attempt is made to cover the entire surface to be examined with body coils for appropriate signal collection. Due to its length, the test is sensitive to motion artefacts, so rapid breathing can cause significant image quality degradation. Proper anaesthesia is therefore essential (Fig. 10.9).

Regarding image treatment, linear and volumetric measurements can be performed on cross-sectional images reconstructed by MRI imaging, similar to images obtained during CT examinations (see Sect. 10.5.2.2).

In contrast with CT, MRI image intensities are not quantitatively standardised. The pixel values depend on the specific MR sequence parameters (e.g. time of repetition, time of echo, flip angle and coil sensitivity) and represent relative signal

Animal: 9121   Gender: male
Date: 23.09.2016   Aquisition Modus: Total Body - Thick

	Total mass	Fat tissue mass	Lean tissue mass	Bone mineral mass	Bone area	Bone mineral density
	g	g	g	g	cm^2	g/cm^2
Head	8005.6	1422.8	6012.9	569.9	232.7	2.45
Arms	9551.9	1884.6	7150.1	517.3	444.5	1.16
Arm Right	4575.9	934.7	3390.0	251.2	216.8	1.16
Arm Left	4976.0	949.9	3760.1	266.1	227.7	1.17
Legs *	16685.8	1691.1	14494.8	499.9	510.0	0.98
Leg Right *	7969.0	824.1	6900.9	244.0	253.4	0.96
Leg Left *	8716.8	867.0	7593.8	255.9	256.6	1.00
Trunk	68187.5	8963.9	58272.1	951.5	911.5	1.04
Trunk Right	34278.9	4555.3	29254.5	469.2	453.5	1.03
Trunk Left	33908.6	4408.7	29017.6	482.4	457.9	1.05
Ribs				286.4	340.9	0.84
Pelvis				211.6	235.9	0.90
Spine				453.4	334.6	1.36
Android	11198.4	1276.6	9830.3	91.5		
Gynoid	8268.4	991.9	6968.3	308.2		
Total	**102430.9**	**13962.4**	**85929.8**	**2538.6**	**2098.8**	**1.21**
Total Right	50976.9	7072.2	42663.4	1241.3	1040.5	1.19
Total Left	51453.9	6890.2	43266.4	1297.3	1058.2	1.23

Fig. 10.8   Example of a data output from a total body DXA acquisition

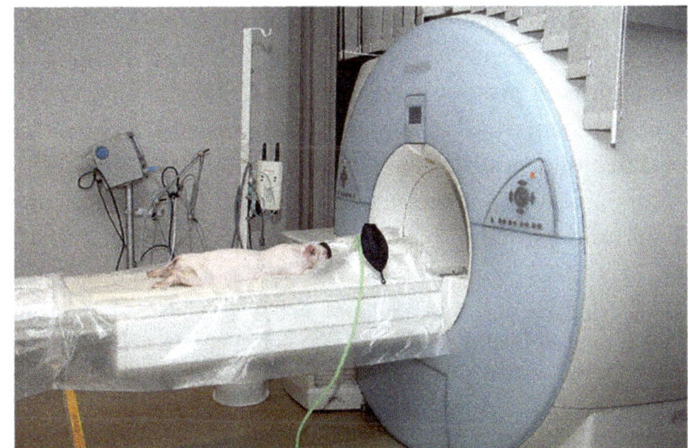

Fig. 10.9   Positioning a pig for MRI scanning before the body coil is placed (Photo credit: Tamás Donkó, Medicopus)

intensities rather than absolute physical properties. Therefore, MRI signal values can vary significantly between sequences and scanner settings. Accordingly, segmentation procedures should take this phenomenon into account.

Results or linear and volumetric measurements are placed in a database and can be treated with specific statistical software.

10.6 Compliance with the 3R Principles

Ethical protocols need to be prepared to carry out the experiments, which will be evaluated by the ethics committee following the established legal procedure in each country. To comply with the 3Rs principles, the animals should be reared in compliance with all welfare standards, for example, with social and environmental enrichment (EOL_0001921) in the pens and proper management. All staff handling the pigs should be trained with regard to stress-free handling, and specific training of pigs should be performed for stress-free immobilisation during ultrasound or for the application of anaesthetics.

The methods described are mainly used in pig nutrition and production studies. The number of pigs required must be calculated in a power analysis. Since the methods are non-invasive and can be performed on live animals, the body composition of the same pig can be repeatedly evaluated during growth. Therefore, a within-individual trial design can be applied, which could result in higher statistical power, needing less animals. In techniques involving manual measurements, staff must receive a thorough training to ascertain high inter-rater-reliability to reduce variance.

10.7 Conclusions

Determination of body composition is necessary to know the characteristics of the animal depending on the treatment applied, sex or genetics, either at one moment of its life or at several moments during its growth, allowing the development of growth curves. The type of information required defines the technology to be used. Standardised protocols are very important to ensure good accuracy and precision of measurements. All the methodologies and technologies described here are non-invasive and non-destructive. Whereas ultrasound allows for the evaluation of the composition at a specific site, DXA, CT and MRI can evaluate the composition of the whole animal.

Acknowledgments The authors would like to thank Dennis Brandborg Nielsen (Danish Technological Institute, DK), Toon Rombouts (University of Ghent, BE) and Sandra Düpjan (FBN Dummerstorf, DE) for reviewing and improving this document.

References

Baulain U (1997) Magnetic resonance imaging for the in vivo determination of body composition in animal science. Comput Electron Agric 17(2) 189–203. https://doi.org/10.1016/S0168-1699(96)01304-X

Carabús A, Gispert M, Font-i-Furnols M (2016) Image analysis techniques to study the composition of live pigs: a review. Span J Agric Res 14(3):1–16

Carabús A, Sainz RD, Oltjen JW et al (2015) Predicting fat, lean and the weights of primal cuts for growing pigs of different genotypes and sexes using computed tomography. J Anim Sci 93:1388–1397

Floradin P, Pomar C, Létourneau-Montminy MP et al (2024) Development of the mineralisation of individual bones and bone regions in replacement gilts according to dietary calcium and phosphorus. Animal 18:101241

Font-i-Furnols M, Luo X, Brun A et al (2020) Computed tomography evaluation of gilt growth performance and carcass quality under feeding restriction and compensatory growth effects on sensory quality of pork. Livest Sci 237:104023

Gonzalo E, Létourneau-Montminy MP, Narcy A et al (2018) Consequences of dietary calcium and phosphorus depletion and repletion feeding sequences on growth performance and body composition of growing pigs. Animal 12:1165–1173

Heurtault J, Hiscocks S, Létourneau-Montminy MP et al (2024a) Dynamics of bone mineralization in primiparous sows as a function of dietary phosphorus and calcium during lactation. Animal 18:101130

Heurtault J, Maïkoff G, Létourneau-Montminy MP et al (2024b) Method: body composition assessment of sows using dual-energy X-ray absorptiometry. Anim Open Space 3:100079

Kasper C, Schlegel P, Ruiz-Ascacibar I et al (2021) Accuracy of predicting chemical body composition of growing pigs using dual-energy X-ray absorptiometry. Animal 15:100307

Kremer PV, Förster M, Scholz AM (2013) Use of magnetic resonance imaging to predict the body composition of pigs in vivo. Animal 7:879–884

Lucas D, Brun A, Gispert M et al (2017) Relationship between pig carcass characteristics measured in live pigs or carcasses with Piglog, fat-o-Meat'er and computed tomography. Livest Sci 197:88–95

Zomeño C, Gispert M, Brun A et al (2016) Predicting the carcass chemical composition and describing its growth in live pigs of different sexes using computed tomography. Animal 10(1):172–181

Image Analysis for the Assessment of Pig Carcass Composition

Maria Font-i-Furnols, Tamás Donkó, Patrick Schlegel, Ludovic Brossard, Marina Gispert, Albert Brun, and Claudia Kasper

Abstract

Knowledge of carcass quality parameters and composition is crucial to determining the value of the carcass and informing decisions regarding its processing. The use of non-destructive technologies for this purpose is essential to avoid the destruction of carcasses and cuts and to reduce waste. This can be achieved through linear measurements, either obtained manually or using semi–/fully automatic devices at the slaughter line. These devices need to be calibrated, and cutting and dissection are used as references for this purpose. This chapter aims to describe the most common carcass quality parameters for characterising carcass composition. Additionally, a list of different aspects that should be considered for device checking and carcass preparation before measuring is provided, as well as the types of measurements possible with each device.

Keywords

Non-destructive · Carcass grading · Dissection · Fat thickness · Lean yield

M. Font-i-Furnols (✉) · M. Gispert · A. Brun
IRTA-Food Quality and Technology, Monells, Catalonia, Spain
e-mail: maria.font@irta.cat

T. Donkó
Medicopus, Kaposvar, Hungary

P. Schlegel
Swine Research Group, Agroscope, Posieux, Switzerland

L. Brossard
PEGASE, INRAE, Institut Agro, Saint Gilles, France

C. Kasper
Animal GenoPhenomics Group, Agroscope, Posieux, Switzerland

L. R. Moscovice et al. (eds.), *Standard Operating Procedures for Better Pig Research (SOPig)*, https://doi.org/10.1007/978-3-032-24899-2_11

11.1 Introduction

Carcass classification is needed to ensure market transparency. The composition of the carcass (ATOL_0005516) is used to determine its lean meat percentage and commercial value, which is a relevant aspect for pig producers and pork processors. Carcass classification depends on the definition of carcass and lean yield, which vary by country (Čandek-Potokar et al. 2024; Marcoux et al. 2007). Today, the different grading systems refer to the overall lean content of the carcasses, not taking into account the composition of the different primal cuts. Apart from lean content, several fat (ATOL_0001517) and muscle thicknesses are also used to classify carcasses according to tissue composition. Several classification methods are available for grading pig carcasses (Pomar et al. 2008; Čandek-Potokar et al. 2024). Carcass information is also relevant in nutritional, genetic and breeding studies. For instance, this information can be used to define nutrient requirements for achieving the desired tissue composition (lean, fat or bone tissues), for genetic improvement or breeding, and to evaluate the effect of castration, immunocastration or other treatments that change the body composition. It can be determined destructively by cutting and dissecting the carcass or non-destructively by using computed tomography (CT) (Olsen et al. 2017; Font-i-Furnols et al. 2021), dual X-ray absorptiometry (DXA) (Mercier et al. 2006) or nuclear magnetic resonance imaging (MRI) technology (Monziols et al. 2006). These technologies can estimate either the chemical or physical composition of the tissues (after a calibration/validation process) or their anatomical distribution.

CT uses X-rays, which are attenuated to varying degrees as they pass through the carcass, depending on tissue density. The CT scan generates a discrete 3D volume composed of numerous voxels (3D pixels). The size of each voxel is determined by the scanner settings, such as the matrix resolution, field of view and slice thickness. Each voxel is assigned a Hounsfield Unit (HU) value, based on the HU scale (e.g., air = -1000 HU, water = 0 HU). In the carcasses, fat has attenuation values between approximately -1 and -160 HU, and lean (muscle) has values between approximately 0 and $+120$ HU. This produces an initial radiographic image (scout or surview) used to select the region to be scanned—either the whole carcass or just a specific region. CT is an authorised reference method in the EU for calibrating pig carcass classification devices "on the condition that satisfactory comparative dissection results are provided" (Regulation (EU) 2017/1182).

Similar to CT, DXA relies on X-ray emissions that are attenuated as they pass through the carcass, depending on the density of the tissues, i.e. lean, fat and bone. DXA produces 2D images that differentiate between these three tissue types. It can be used to scan entire carcasses or specific cuts. The output includes measurements such as lean tissue weight, fat tissue weight, bone mineral content, skeletal surface area and areal bone mineral density. The sum of lean, fat and bone mineral masses equals the total body mass. Within the DXA image, regions of interest (ROIs) are defined, and results are reported separately for each ROI as well as for the total scanned area.

As with CT and DXA, MRI has the advantage of being non-invasive. MRI is based on the principle that atomic nuclei with an odd number of protons and/or neutrons absorb and reemit radio waves when placed in a magnetic field. Hydrogen, the most abundant element in the body, has the simplest nucleus, a single proton; therefore, it exhibits a magnetic moment. In contrast, elements like carbon (C) and oxygen (O) do not possess a magnetic moment due to their even numbers of protons and neutrons. The presence of a magnetic moment in hydrogen nuclei enables them to emit a much stronger signal. This property makes hydrogen an especially suitable isotope for imaging applications. To align the randomly oriented protons in the body, a uniform magnetic field is required. The frequency at which nuclei resonate, known as the Larmor frequency, is directly proportional to the strength of the magnetic field. The response signal of the protons varies based on the tissue in which they are embedded because of the bonded atoms affecting the electromagnetic field. This phenomenon can be leveraged to distinguish protons in different tissues. After the transmission of a radio frequency (RF) pulse, tissue protons release the absorbed energy, generating a weak RF signal, used to reconstruct cross-sectional images of the body (Baulain 1997).

11.2   Goal and Scope of the Procedure

The goal of this procedure is to describe the most common carcass quality parameters used to characterise the composition of the carcasses, mainly in terms of morphometric measurements, fat (ATOL_0005521), and lean percentage and conformation (ATOL_0005555). Additionally, a list of different aspects that should be taken into account when determining carcass quality with non-invasive technologies (CT, DXA and MRI) is provided. The procedure includes the individual carcasses of the slaughtered pigs. Usually, the weight is the typical carcass weight (ATOL_0001057) (between 60 and 120 kg), but the procedure can also be applied to carcasses of lower or higher weight.

11.3   Materials and Equipment

11.3.1 Ruler, Calliper and Tape

Fat and muscle thickness can be measured at the midline using a ruler or a calliper. The length of the ruler should be a minimum of 10 cm, with a precision of 1 mm. Some carcass measurements, such as loin and carcass length (ATOL_0001543), require a measuring tape, ideally made of metal. The length should be a minimum of 1.5 m, with a precision of 1 mm. The ruler, calliper and tape used to take lineal measurements on the carcass must be metallic to avoid disturbances due to wear and tear and the temperature.

11.3.2 Knives and Saw

To cut and dissect the carcasses, several knives and a saw (preferably a circular saw) are needed.

11.3.3 Scales

Weighing the carcass requires a monorail weighing scale (if possible) with a precision of a minimum of 500 g. A scale that measures up to 20–30 kg with a precision of a minimum of 1 g is also needed to weigh the cuts and tissues.

11.3.4 Conformation Pattern

A pattern or photographic model is used to determine the conformation of the carcasses. It would allow us to classify carcasses in several classes according to their conformation. An example of poorly and highly conformed carcasses is presented in Fig. 11.1.

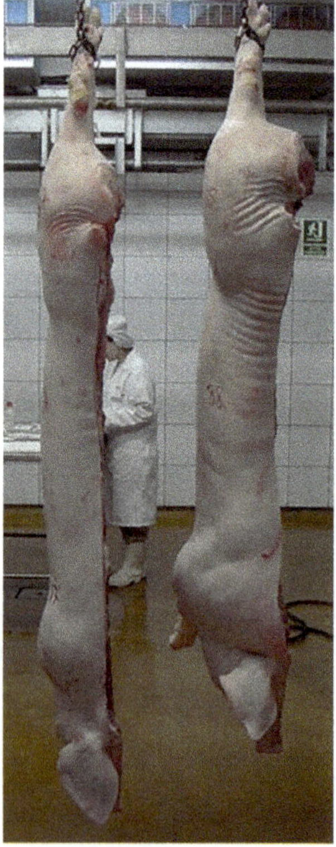

Fig. 11.1 Example of poorly (left) and highly (right) conformed carcasses

11.3.5 Carcass Classification Devices

Lean meat percentage and/or fat and muscle thickness can be measured using several classification devices. These devices can be manual, semi-automatic or fully automatic and can be based on different technologies, such as vision, reflectance, ultrasound or magnetic induction.

- Regarding the level of automatisation:
    - Manual devices require both the measurement and subsequent reading of the results to be carried out by a trained operator (e.g. a ruler).
    - Semi-automatic devices involve manual measurement by a trained operator, while the reading of the results is automated. Examples include the Fat-O-Meat'er and Ultrafom (Frontmatec AS, DK), Hennessy Grading Probe (Hennessy Grading System LTD, NZ), Capteur Gras Maigre (Sydel, PL) and Optiscan TP (Classpro GmbH, DE).
    - Fully automatic devices operate without the need for an operator for either measurement or result readout. Examples include the Autofom (Frontmatec AS, DK), VCS2000 (e + V Technology GmbH, DE), CSB-Image-meater (CSB System, DE), Tobec (Meat Quality Inc., US) and gmSCAN (gmsteel, IR).
- Regarding the technologies, the most commonly used are:
    - Reflectance: Fat-O-Meat'er, Hennessy Grading Probe, Capteur Gras Maigre.
    - Ultrasound: Autofom, Ultrafom.
    - Vision: VCS2000, CSB-Image-Meater, Optiscan-TP.
    - Magnetic induction: gmSCAN.

Other technologies, including computed tomography, dual X-ray absorptiometry and nuclear magnetic resonance imaging, can also be used for carcass assessment (see Chap. 10, Sect. 10.3.1). Within the EU, each country has its own approved devices. The choice of one or another depends on the capacity and speed of the slaughter plant as well as the economic resources.

11.4 Prerequisites and Preparation

11.4.1 Scales

The scales used for carcass measurements must comply with the regulations of metrological control (2014/32/EU). They must be calibrated internally with an adequate and standardised pattern before each use. Each day, before use, they must be verified using a standard mass (i.e. a piece with a specific weight). The permissible error margin must be stated before each trial.

11.4.2 Carcass Classification Devices

As a prerequisite, the devices used in EU slaughter plants should have official prediction equations approved by the EU for each country, following the required protocol (official carcass presentation, dissection or CT scans of a minimum of 120 carcasses and error of prediction less than 2.5% root mean square error or prediction-RMSEP).

It is very important to comply with the regulations currently in force, including any updates that may have been issued. Current relevant regulations include:

- Regulation (EU) No 1308/2013 of the European Parliament and of the Council of 17 December 2013 establishing a common organisation of the markets in agricultural products and repealing Council Regulations (EEC) No 922/72, (EEC) No 234/79, (EC) No 1037/2001 and (EC) No 1234/2007.
- Commission Delegated Regulation (EU) 2017/1182 of 20 April 2017 supplementing Regulation (EU) No. 1308/2013 of the European Parliament and of the Council as regards the Union scales for the classification of beef, pig and sheep carcasses and as regards the reporting of the market prices of certain categories of carcasses and live animals.
- Commission Implementing Regulation (EU) 2017/1184 of 20 April 2018 laying down rules for the application of Regulation (U) N. 1308/2013 of the European Parliament and of the Council as regards the Union scales for the classification of beef, pig and sheep carcasses as regards the reporting of the market prices of certain categories of carcasses and live animals.

Each country has its own approved devices and prediction equations, and there is no common equation for all countries. Outside of the EU, legal requirements, if any, should be followed.

Regarding preparation, each classification device requires daily verification before use. The verification process varies depending on the device, and the manufacturer's instructions must be followed for proper use. In Fig. 11.2, different types of verification "phantoms" or pieces are presented.

For manual and semi-automatic devices, the standardisation of their use is as important as the verification of the devices. The operators should be monitored periodically (in accordance with the established protocol) to ensure that measurements are taken correctly. This can be achieved either by using a needle to determine the

Fig. 11.2 Different types of patterns or "phantoms" for verifying carcass classification devices: (**a**) Pig plastic pattern for Fat-O-Meat'er (Frontmatec AS, DK), (**b**) Block for Autofom (Frontmatec AS, DK)

direction and location of the measurement (Fig. 11.3) or by taking measurements from several trained operators and comparing the results. When the devices take the measures in the carcass midline, it is important that the carcass is split correctly.

11.4.3 CT/DXA/MRI

All equipment and installations must comply with all legal requirements relating to the safety of personnel, such as shielding, armouring, emergency plans and emergency stops (Table 11.1).

11.4.4 Carcass Preparation and Cutting

After bleeding, dehairing and eviscerating, the carcass is divided into two halves. Usually, the left half-carcass is used for quality measurements. It is important to

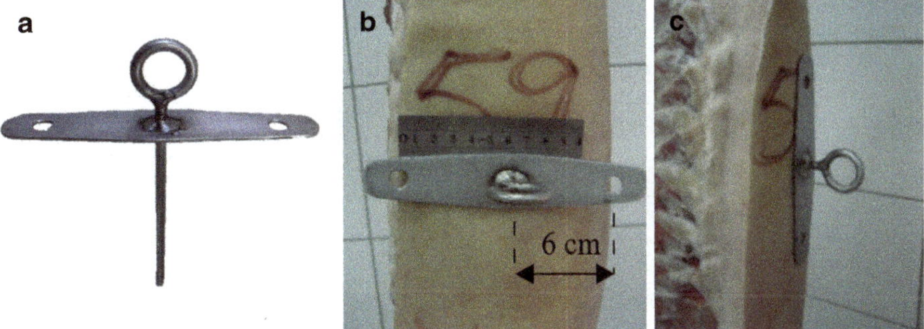

Fig. 11.3 (**a**) Needle to be inserted in the place of measurement of the carcass with the Fat-O-Meat'er (Frontmatec AS, DK) to control the correctness of the measure in terms of (**b**) Location and (**c**) Direction of the measure

Table 11.1 Aspects to consider regarding the devices and the general scanning procedure of the installations/devices

General procedure	CT	DXA	MRI
Legal requirements to work (armour, etc.)[a]	X	X	X
Calibration of the device previous to the measurement	X	X	X
Individual radiation badge to control radiation[a]	X	X	
Distance to control radiation[a]	(X)	X	
Operational room with armour during evaluation	X		
Room and material with no metallic pieces			X
Distance of the material from the magnetic field with cardiac pacemaker, defibrillators, insulin pump, implants made of metal, hearing instruments, dosage devices for medication, etc.			X

[a]May differ from country to country

ensure that carcasses are split correctly to avoid biases between the two halves and to enable correct measurements at the carcass midline. According to EU legislation (Regulation (EU) No 1308/2013), carcass presentation is defined as follows: "Carcasses shall be presented without tongue, bristles, hooves, genital organs, flare fat, kidneys and diaphragm". Any defect in carcass presentation, such as poor splitting, must be recorded. Alternative presentations can be used, but coefficients are then needed to correct for differences in carcass weight. A checklist of aspects to be considered when preparing the protocol for carcass presentation is presented in Table 11.2. Carcasses that meet the required definition are then scanned and/or dissected. This should be done after the carcasses have been thoroughly chilled to a core temperature of 4–7 °C within 48 h post-mortem.

If the objective is to evaluate the composition of the pieces, the carcasses should be cut. This can be done by following either the EU reference method or a commercial cut (Fig. 11.4). In Canada, the cutting is different from the EU since carcasses are cut into pieces (see Marcoux et al. 2007).A checklist of aspects to be considered when preparing the protocol for the presentation of cuts is shown in Table 11.2.

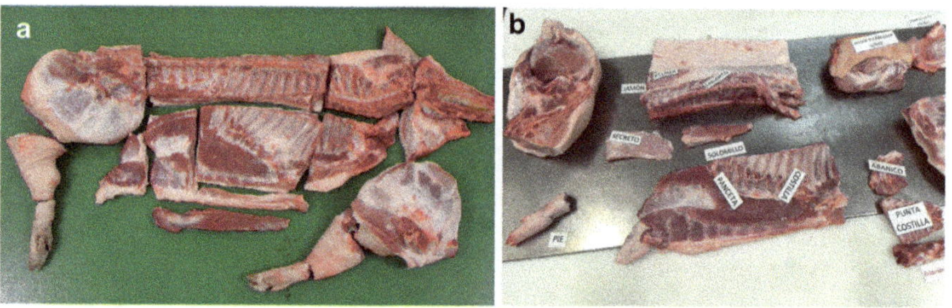

Fig. 11.4 Reference EU cutting (**a**) and an example of commercial cutting (**b**)

Table 11.2 Aspects to be considered in the preparation of the carcasses and cuts

Carcass and cut preparation
Check if the carcass is correctly split
Carcass definition If EU definition of carcass: Without flare, flat, kidneys, diaphragm, etc. If others: Adapt to each objective (i.e. include half of the head if information about the full animal is targeted)
If the carcass is scanned and cut into pieces, decide on the cutting procedure (EU reference or commercial procedure). However, caution should be taken when scanning pieces, because, at least with CT, a scan of ½ carcass and ½ carcass in pieces does not add to the same value due to mixed voxel (known as partial volume effect)
If the carcass is scanned following the EU reference procedure (Reg.1182/2017), the head (except for the cheek), and feet need not to be scanned

11.4.5 Carcass and Cut Positioning for Scanning

When scanning carcasses with CT, DXA and MRI, the position of the carcass is important. Table 11.3 provides a checklist of different aspects that need to be taken into consideration when positioning carcasses (and cuts) in the scanner for evaluation.

11.4.6 Definition of Scanning Protocol

The scanning protocol needs to be defined based on previous experience, pre-study experimental trials, the functionality of the device, the objective of the evaluation or the literature review carried out. Some aspects to be considered are presented in Table 11.4.

Table 11.3 Aspects relative to the position of the carcass (cut) for scanning

Scanning positioning
Situation of the carcasses (cuts) skin side up or skin side down
If the carcass is scanned and cut into pieces, consider scanning all the pieces together or cut by cut. If all pieces are scanned together, consider the way to position the cuts on the table of the device because the orientation of the pieces might have an effect
Warm or cold carcass scanning
Scanning limitation of length and width. Check for limitations in advance, and if there are limitations, remove the parts of the carcass that cannot be fitted in the field of view of the device (e.g. hind shank)
In CT and MRI, removed pieces can be placed on top of the carcass, ideally separated with an X-ray transparent piece (e.g. insulation board) to facilitate later image processing
In DXA, it is necessary to run a second scan if not all pieces can be positioned separately on the table
Remove all the metallic pieces (ear tag, etc.)
In MRI, the pieces to be measured must be placed in the measuring field so that the coils cover them and can move freely in the measuring tunnel

Table 11.4 Aspects to be considered for the scanning or evaluation of the carcasses and cuts

Scanning protocol
Definition of the region to be scanned (specific anatomical area) or scan the full carcass (cut)
Definition of the acquisition parameters (depends on the device capacity and the objectives of the evaluation)
Definition of the circumstances during the MRI acquisition process (mainly core and room temperature)

11.5   Description of the Procedure

11.5.1 Weights of the Carcass and Cuts

The weight of the carcass, prepared according to the pre-established presentation, is recorded either manually or automatically.

11.5.2 Conformation

Visual confirmation is determined by trained technicians/operators following the photographic model.

11.5.3 Midline Tape and Ruler or Calliper Measurements

The tape is used to take the following measurements, as described below:

- Carcass length (cm): Carcass length measured from the anterior edge of the *symphysis pubic* to the recess of the first rib (Fig. 11.5a).
- Loin length (cm): Length from the atlas to the first lumbar vertebra (Fig. 11.5a).

The ruler or calliper is used to measure in the midline. Some of the measurements are the following:

- Fat thickness at the last rib level (P2) mm: minimum fat thickness (including skin, ATOL_0005636) measured perpendicularly to the skin at the level of the last rib.
- Fat thickness between the third and fourth last rib level (mm): Minimum fat thickness (including skin) measured perpendicularly to the skin at the level between the third and fourth last rib.
- Fat thickness at the first lumbar vertebra level (mm): minimum fat thickness (including skin) measured perpendicularly to the skin at the level of the first lumbar vertebra.
- ZP ("Zwei-Punkt-Messverfahren") fat thickness (mm): the shortest measurement of fat plus skin thickness over the *gluteus medius* muscle (Fig. 11.5b).
- ZP muscle thickness (mm): defined as the distance from the vertebral channel to the cranial end of the *gluteus medius* muscle (Fig. 11.5c).

11.5.4 Measure with Carcass Classification Devices

The measurement can be automatic, semi-automatic or fully manual, depending on the device available (see Sect. 11.3.5). If semi-automatic devices are used (i.e. Fat-O-Meat'er, Hennessy Grading Probe, Capteur Gras Maigre, Optiscan TP), the measurement should be performed at the position according to the purpose of the

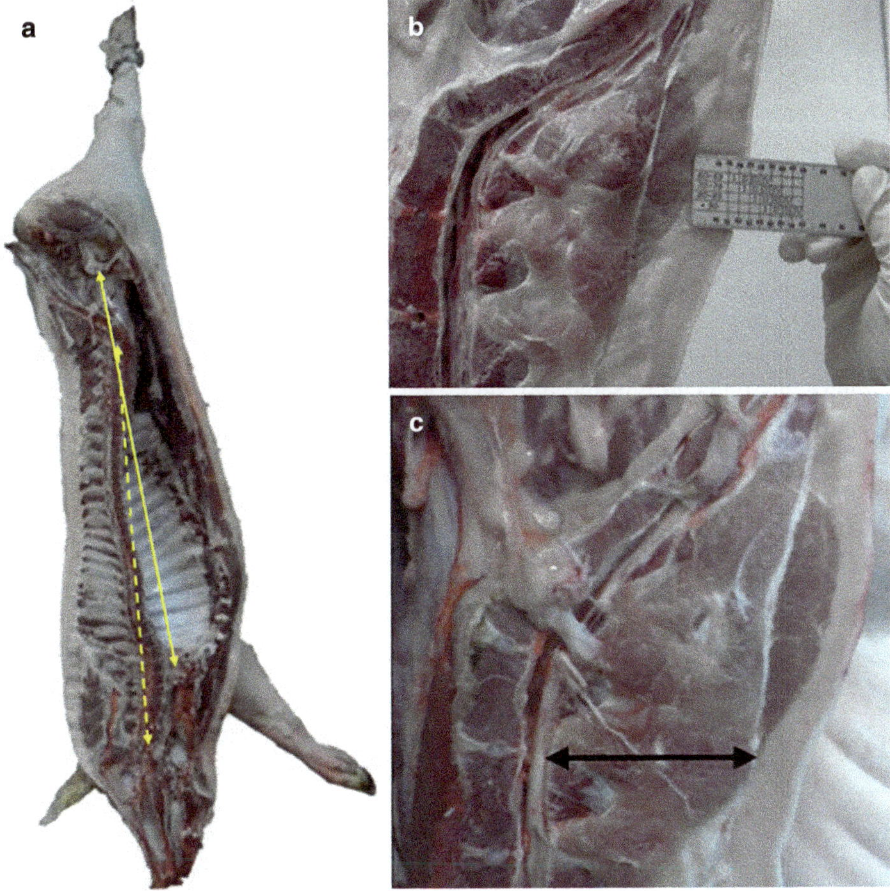

Fig. 11.5 Carcass quality measurements: (**a**) carcass (continuous line) and loin (discontinuous line) length, (**b**) ZP ("Zwei-Punkt-Messverfahren") fat thickness, and (**c**) ZP muscle thickness

measure. For instance, if fat and muscle thickness are to be used to measure lean meat percentage, measurements are taken at the site defined by the prediction equation used for the determination.

The results obtained with these classification devices can be:

- Fat and muscle thickness at specific locations: all the devices.
- Fat and muscle areas: fully automatic devices based on vision image analysis.
- Other types of measurements (minimum thickness, maximum thickness, etc.): automatic devices.
- Carcass lean meat percentage: predicted by all the devices according to the official calibration equation and/or an internal equation provided by the manufacturer.
- Cuts lean meat percentage: predicted from the carcass evaluation if the calibration equations are previously obtained.

11.5.5 Carcass and Cuts Dissection and Lean Meat Percentage Determination

The different cuts obtained from the carcass can be dissected to separate the different tissues and consecutively weighed to obtain the lean meat percentage. There are two types of dissection:

- Total (full) dissection: dissection of all cuts. Separation of the skin, subcutaneous fat (ATOL_0005562), intermuscular fat, bones and lean. The feet and the head, excluding the cheek, are not dissected.
- Partial (simplified) dissection: dissection of the four main cuts, that is, ham, loin, shoulder and belly. Separation of skin, subcutaneous fat, intermuscular fat, bones and lean. Skin and subcutaneous fat do not need to be separated.

An example of the dissection sheet that can be used to record the data is presented in Fig. 11.6.

Once the dissection is completed, the carcass lean meat percentage (LMP) is calculated as follows if the total dissection has been performed:

$$LMP = 100 \times \frac{Weight\ of\ the\ lean\ of\ all\ the\ pieces}{Carcass\ weight}$$

For simplified dissection, the LMP is calculated as follows:

$$LMP = \partial \times \frac{Weight\ of\ the\ lean\ of\ ham, shoulder, belly, loin + weight\ of\ tenderloin}{Weight\ of\ ham, shoulder, belly, loin + tenderloin}$$

CUTS	WEIGHT	SKIN	SUBC. FAT	INTER. FAT	BONES	LEAN
HAM						
LOIN						
SHOULDER						
BELLY						
TENDERLOIN						
HIND SHANK						
NECK						
FRONT SHANK						
VENTRAL PART OF BELLY						
JAWL						
CHECK						
TO VENTRAL PART BELLY						
HAND						
FOOT						
HEAD						

Fig. 11.6 Sample form for simplified dissection (green), full dissection (red + green) or non-dissection (grey). Main cuts are indicated with bold, right-justified text. Note that in simplified or partial dissection, the skin is not separated from subcutaneous fat

where ∂ is a coefficient that should be calculated to allow the LMP obtained from partial dissection to approximate the LMP derived from full carcass dissection (Commission Delegated Regulation (EU) 2017/1182). The lean weight is calculated by subtracting the total weight of the non-lean elements from the initial carcass weight before dissection.

The lean meat percentage content allows to classify carcasses into six groups according to Regulation (EU) No 1308/2013, i.e. Class S ($\geq$ 60% or more), E ($\geq$ 55%), U ($\geq$ 50%), R ($\geq$ 45%), O ($\geq$ 40%) and P ($<$ 40%).

11.5.6 Carcass Measurements with Computed Tomography

To use CT as an official reference method for lean meat percentage determination, comparative dissection results must be provided (Regulation (EU) 2017/1182). For this purpose, CT has to be calibrated against total dissection, that is, the percentage of lean weight with respect to carcass weight. Only the left half-carcass is scanned. According to the legislation (Annex 5, Part A): "If this CT procedure is not calibrated to the total dissection of carcasses, a potential bias to total dissection is corrected based on a subsample that is totally dissected according to the reference method. Only that part of the left half carcass containing lean meat as defined by the total dissection method needs to be scanned, i.e. the feet and the head, except the cheek, need not be scanned". For more detailed information on this dissection, see Sect. 11.5.5.

Furthermore, according to the cited legislation: "The bias correction required for partial dissection or for a CT procedure is based on a representative subsample that includes all combinations of the sample with respect to the stratification factors such as breed, gender or fatness used to select the overall sample. At least 10 carcasses are selected for bias correction. If the slaughter pig population to be sampled has the same characteristics as the population for which partial dissection or a CT procedure has been previously bias corrected, no additional total dissection is required. If a CT procedure is described and is traceable by measurements to total dissection or another, bias-corrected CT procedure, no additional total dissection is required".

To predict lean meat percentage it is necessary to use all the images of the carcass. Thus, from all the images, it is possible:

- To obtain the distribution of volume associated with each Hounsfield value (Fig. 11.7). From this volume, it is possible to predict the lean meat percentage and also the lean weight, fat weight, fat percentage, protein and total fat by applying previously obtained calibration equations. It is also possible to calculate total volume associated with fat, lean or bone according to the Hounsfield value.
- To apply several segmentation techniques to classify the voxels in lean, fat or bone, whether or not the skin and marrow are removed, depending on the objectives of the project (Fig. 11.8).

In addition to calculating lean meat percentage, computed tomography images (typically saved and stored as DICOM image format) can be used to determine

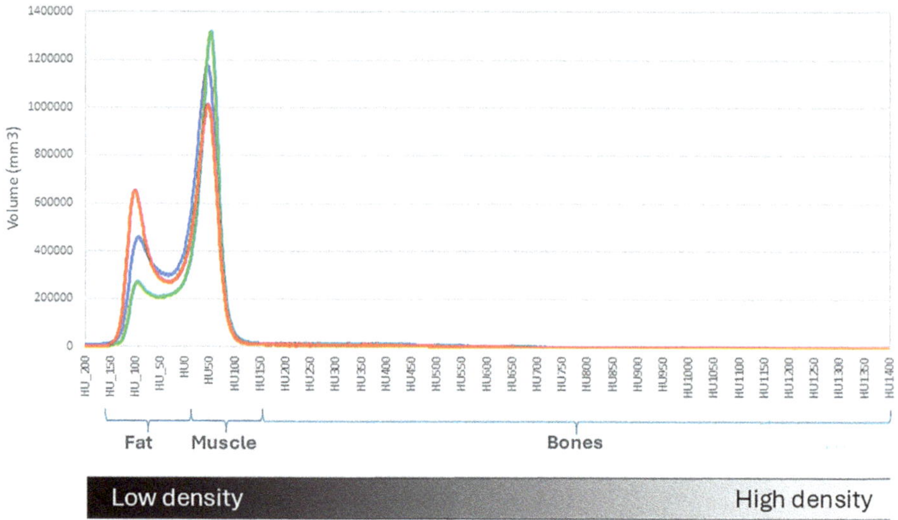

Fig. 11.7 Volume associated with each Hounsfield value of three pigs: a fat pig (red), a medium fat pig (blue) and a lean pig (green)

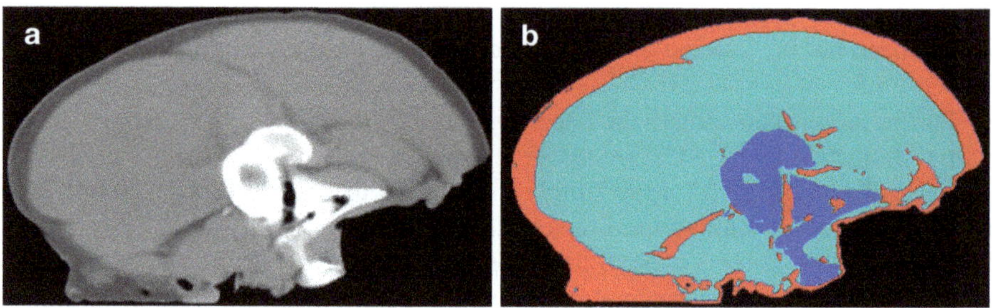

Fig. 11.8 Example of a computed tomography image of the ham region of a pig half carcass before (**a**) and after (**b**) segmentation of the different tissues (red: fat, teal: lean, blue: bones)

several linear measurements and areas in carcasses. Some of the measurements are defined below, but they are not established by law and can be modified depending on the objectives of the project.

- Fat thickness between the third and fourth last rib level (mm): minimum fat thickness (including skin) measured perpendicularly to the skin at the level between the third and fourth last rib and at 60 mm of the midline (Fig. 11.9).
- Loin area of the longissimus muscle determined at the last rib level.
- Loin area of the longissimus muscle determined between the third and fourth last rib levels (Fig. 11.9).

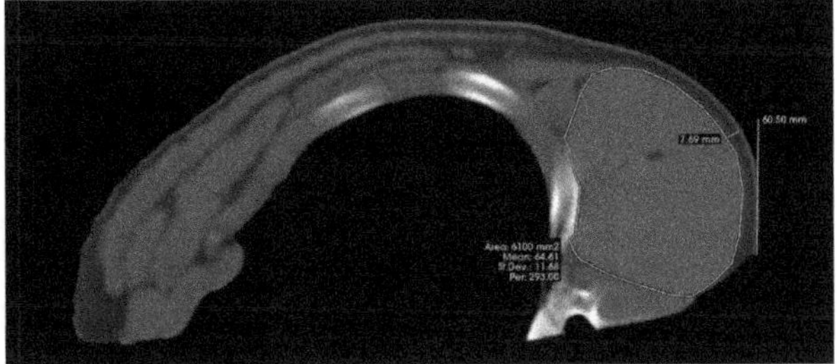

Fig. 11.9 Fat thickness measured at 6 cm of the midline and loin area in the axial computed tomography image image at the level between the third and fourth last ribs

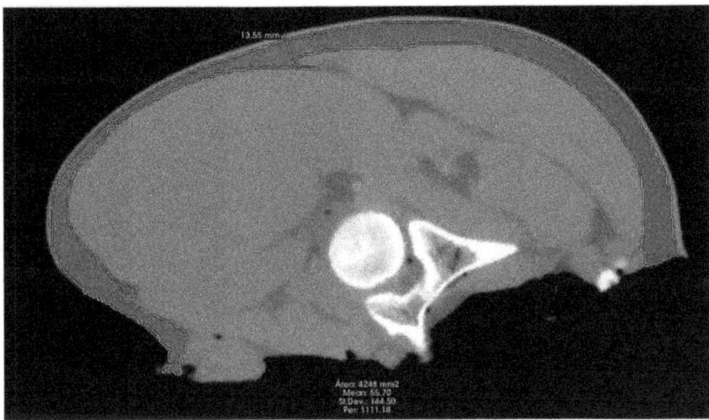

Fig. 11.10 Area and fat thickness measurements in the computed tomography axial image of a pig ham at the level of the joint between the femur and pelvic bones (source: adapted from Font-i-Furnols et al. 2021)

- Fat thickness (excluding skin) measured at the junction between the biceps femoris and tensor fasciae latae muscles in the CT axial image obtained at the level of the joint between the femur and the pelvic bones of the ham region (Fig. 11.10).
- Subcutaneous fat area (excluding skin) measured at the junction between the biceps femoris and tensor fasciae latae muscles in the CT axial image obtained at the level of the joint between the femur and the pelvic bones of the ham region (Fig. 11.10).
- Fat thickness (excluding skin) measured at the junction between the semimembranosus and biceps femoris muscles at the CT axial image obtained just beside the patella in the caudal direction.
- Subcutaneous fat area (excluding skin) measured at the junction between the semimembranosus and biceps femoris muscles at the CT axial image obtained just beside the patella in the caudal direction.

11.5.7 Carcass Composition with Dual X-Ray Absorptiometry

The positions of the half carcass and carcass cuts in a DXA device are shown in Figs. 11.11 and 11.12, respectively.

Depending on the DXA device, several acquisition modes for the complete body are available. For example, GE Healthcare iDXA has the following modes:

- Total body—thick: mean height of >25 cm.

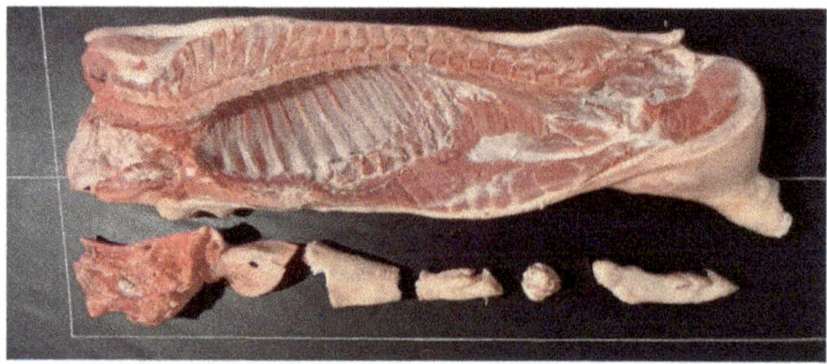

Fig. 11.11 Position of a half carcass on a DXA device

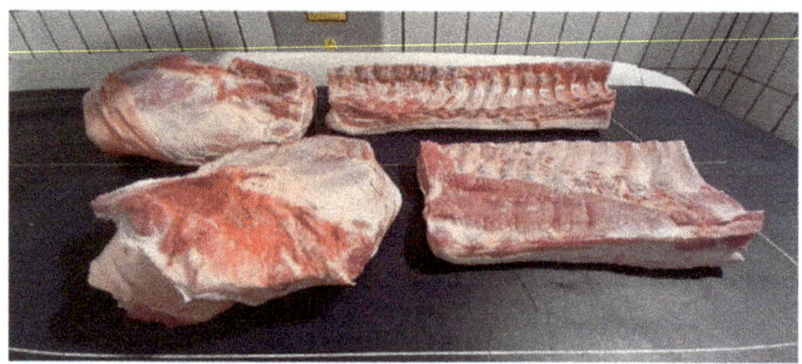

Fig. 11.12 Position of carcass cuts on a DXA device

- Total body—standard: mean height of 16–25 cm.
- Total body—thin: mean height of <16 cm.
- Small animal body—large: > 20 kg.
- Small animal body—medium: 2–20 kg.
- Small animal body—small: < 2 kg.

At Agroscope, all carcasses are scanned using total body (thick mode), defined as the standard mode for live animal scans. In addition, total body (standard) or small animal body (large mode) settings may be used for half carcasses from pigs with lower BW.

Although DXA acquisition modes can be adapted to carcass size, the calculation of tissue masses varies between modes. To ensure comparability of results across carcasses of different sizes within and between experiments, and with live scans, it is advisable to use a single acquisition mode (typically the one suitable for the heaviest carcasses) for all scans, even if this deviates from the manufacturer's recommendations for smaller body sizes.

The image treatment can be done at any time after the acquisition and consists of the following:

1. Adapting the placement of the ROI. As the equations to determine lean tissue, fat tissue and bone mineral masses from the X-ray signals depend on the ROI, the correct placement of the ROI on the body is relevant. As the ROI are designed for humans in the supine position, a correct placement of the ROI is not possible for half carcasses, representing a lateral position. Therefore, all the lines delimiting the ROI are moved to the right side of the scan image (Fig. 11.13, left). As a result, the half-carcass is recognised and calculated within the ROI of the "right

Fig. 11.13 Example of regions of interest (ROIs) studied with DXA in the entire half carcass (a) and in the cut half carcass (b)

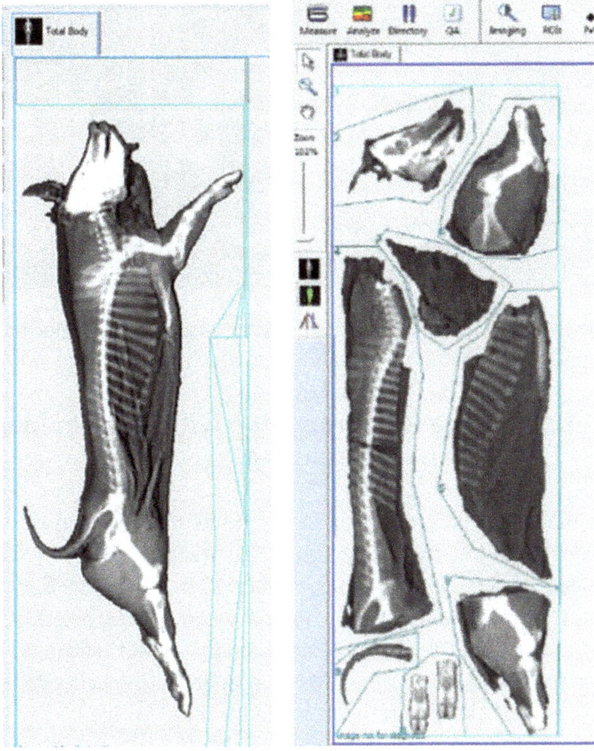

Fig. 11.14 Example of data obtained from DXA scanning of a half carcass in cuts

Animal: 9121 Gender: male
Date: 23.09.2016 Aquisition Modus: Total Body · Thick

	Total mass	Fat tissue mass	Lean tissue mass	Bone mineral mass	Bone area	Bone mineral density
	g	g	g	g	cm²	g/cm²
Total half-carcas	39789.8	7699.4	30693.1	1397.3	1950.2	0.72
Half-head	2661.3	599.2	1764.0	298.1	331.1	0.90
Neck	1958.2	719.6	1238.6	0.0	0.0	0.00
Back	13248.9	2407.8	10453.1	388.0	622.9	0.62
Belly	6647.3	1663.5	4901.0	82.9	274.4	0.30
Shoulder	6200.2	867.3	5095.8	237.0	286.5	0.83
Ham	7933.6	984.8	6703.9	244.9	225.8	1.08
Feet	900.9	377.8	390.3	132.8	166.9	0.80
Tail	239.4	79.3	146.5	13.6	42.6	0.32

arm". In the case of scans of cuts, a customised ROI can be defined (after moving the standard ROI, as indicated above) for each cut to have the results per piece (Fig. 11.13, right).

2. Checking the correct attribution to body tissues. Any parts that are not from the animal should be labelled as "artefacts".

Depending on the device, data can be exported as a data file, which is an advantage for research purposes, as the focus is on the population of all animals scanned rather than on one single "patient". With GE Healthcare iDXA, the exported results are arranged by region of interest (customized), as illustrated by an example of half carcass cuts shown in Fig. 11.14.

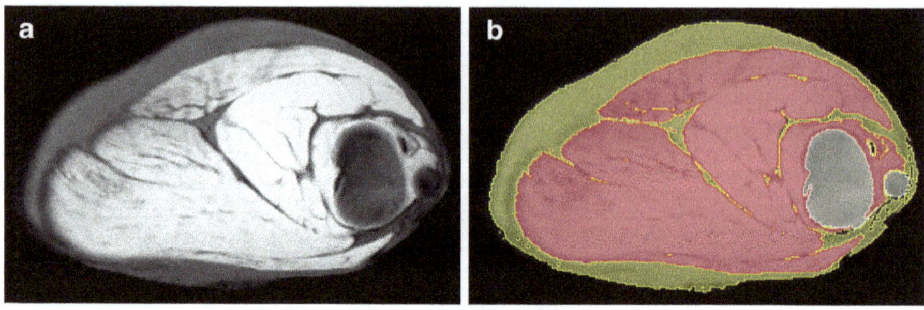

Fig. 11.15 Example of **a**) native MRI image and **b**) segmented lean (red), bone (gray) and fat tissues (yellow)

11.5.8 Carcass Composition with Nuclear Magnetic Resonance Imaging

Linear and volumetric measurements can be performed on cross-sectional images reconstructed by MRI, similar to images obtained during CT examinations (see Sect. 11.5.6). However, as explained in Chap. 10, Sect. 10.5.2, MRI image intensities are not quantitatively standardised. Therefore, MRI signal values can vary significantly between sequences and scanner settings. Accordingly, segmentation procedures should take this phenomenon into account. An example of the segmentation of a MRI image is presented in Fig. 11.15.

11.6 Compliance with the 3R Principles

Dissecting the carcass with a knife causes damage and reduces its overall value, although the separated tissues can still be recycled and repurposed for other uses (if the cooling chain has been maintained and all the permissions are available). However, dissection is needed to calibrate non-destructive devices. Another definition of lean percentage that is commonly used in the United States is fat-free lean, where the lean percentage is determined chemically, hence destructively. In this case, nothing can be reused or recycled.

Apart from dissection, all the linear, area and compositional measurements explained above are non-destructive; thus, it is possible to use the carcasses for human consumption after the measurements without losing value. Regarding the reduction, the number of carcasses to be evaluated needs to be calculated in a power analysis to have a robust model or according to what the legislation establishes. This also applies to the use of CT, DXA and MRI for carcass composition determination.

Staff training is essential to ensure high inter-rater reliability, particularly in carcass preparation, placement and non-automated measurements, which helps minimise variance and, consequently, reduces the required sample size.

11.7   Conclusions

The EU legislation and national legislation require carcass classification, that is, the determination of lean meat percentage, to increase market transparency. Carcass composition can be determined in many ways using a wide range of available methods, varying from simple ruler measurements to complex image analysis. The information gathered can be used to assess animal performance and perform carcass classification (determination of lean meat percentage). The calibration of abattoir online devices for the determination of lean meat percentage needs a reference, which, in the EU, is the total dissection of carcasses. However, virtual dissection with non-invasive and non-destructive devices, such as CT (and MRI), is also allowed if satisfactory comparative dissection results are provided. In any case, it is very important to standardise the procedure to obtain accurate and precise information.

Acknowledgements The authors would like to thank Dennis Brandborg Nielsen (Danish Technological Institute, DK) and Toon Rombouts (University of Ghent, BE) for reviewing and improving this document.

References

Baulain U (1997) Magnetic resonance imaging for the in vivo determination of body composition in animal science. Comput Electron Agric 17:189–203

Čandek-Potokar M, Lebret B, Gispert M et al (2024) Challenges and future perspectives for the European grading of pig carcasses—a quality view. Meat Sci 208:109309

Commission Decision of 19 December 2008 authorising methods for grading pig carcases in Spain (notified under document number C (2008) 8477) (Only the Spanish text is authentic) (2009/11/EC)

Commission Delegated Regulation (EU) 2017/1182 of 20′ April 2017 supplementing Regulation (EU) No. 1308/2013 of the European Parliament and of the Council as regards the Union scales for the classification of beef, pig and sheep carcasses and as regards the reporting of the market prices of certain categories of carcasses and live animals

Commission Implementing Decision (EU) 2018/114 of 16 January 2018 amending Decision 2009/11/EC authorizing methods for grading pig carcasses in Spain

Commission Implementing Regulation (EU) 2017/1184 of 20 April 2018 laying down rules for the application of the Regulation (EU) N. 1308/2013 of the European Parliament and of the Council as regards the Union scales for the classification of beef, pig and sheep carcasses as regards the reporting of the market prices of certain categories of carcasses and live animals

Decision Commission Implementing Decision (EU) 2018/1521 of 10 October 2018 amending Decision 2009/11/EC authorizing methods for grading pig carcasses in Spain

Decision Commission Implementing Decision (EU) 2020/113 of 23 January 2020 amending Decision 2009/11/EC authorizing methods for grading pig carcasses in Spain

Font-i-Furnols M, García-Gudiño J, Izquierdo M et al (2021) Non-destructive evaluation of carcass and ham traits and meat quality assessment applied to early and late immunocastrated Iberian pigs. Animal 15(4):100189

Marcoux M, Pomar C, Faucitano L et al (2007) The relationship between different pork carcass lean yield definitions and the market carcass value. Meat Sci 75(1):94–102

Mercier J, Pomar C, Marcoux M et al (2006) The use of dual-energy X-ray absorptiometry to estimate the dissected composition of lamb carcasses. Meat Sci 73:249–257

Monziols M, Collewet G, Bonneau M et al (2006) Quantification of muscle, subcutaneous fat and intermuscular fat in pig carcasses and cuts by magnetic resonance imaging. Meat Sci 72:146–154

Olsen EV, Christensen LB, Nielsen DB (2017) A review of computed tomography and manual dissection for calibration of devices for pig carcass classification - evaluation of uncertainty. Meat Sci 123:35–44

Pomar C, Marcoux M, Gispert M et al (2008) Determining the lean meat content of pork carcasses. In: Kerry J, Ledward D (eds) Improving the sensory and nutritional quality of fresh meat (capítulo 21). Woodhead Publishing Limited, Cambridge, pp 493–518

Regulation (EU) No 1308/2013 of the European Parliament and of the Council of 17 December 2013 establishing a common organization of the markets in agricultural products and repealing Council Regulations (EEC)

The Value of Experimental Animal Research

12

Jaap van Milgen

Abstract

In this last chapter of the book, some information about the PIGWEB project (i.e. the umbrella under which the SOPs were developed) is given, and I will share some personal thoughts on the value(s) of our current research activities. Looking back on how experimental animal and human research was done in the past should help us realize how future generations will look upon the activities that we carry out now. Most of us have been trained in a way that if we have a research question, the answer is to be found via an experiment. We can do this because we have access to these animals. This is very different from human research that relies to a great extent on indicators, because they have little or no direct access to human subjects. Human research methods should be a source of inspiration for animal researchers: we can view the focus on indirect indicators in human research as a model for what the future of pig research may look like.

Keywords

Research infrastructures · Experimental animal research · Ethics · Standard operating procedures

12.1 Introduction

In this last chapter of the book, I provide some information about the PIGWEB project (i.e. the umbrella under which the SOPs were developed) and share some personal thoughts on the value(s) of our current research activities.

PIGWEB was a research infrastructure project of the Horizon 2020 work program of the European Commission. The project was carried out between 2021 and 2026, involved 16 partners from 11 countries, and had a budget of five million euros. PIGWEB was a so-called "starting community", with the ambition to bring

J. van Milgen (✉)
PEGASE, INRAE, Institut Agro, Saint Gilles, France
e-mail: jaap.vanmilgen@inrae.fr

L. R. Moscovice et al. (eds.), *Standard Operating Procedures for Better Pig Research (SOPig)*, https://doi.org/10.1007/978-3-032-24899-2_12

together and open up key national and regional research infrastructures. Starting communities in a research infrastructure project were expected to include:

Networking activities (NA), to foster a culture of cooperation between research infrastructures, scientific communities, industries, and other stakeholders as appropriate, and to help develop a more efficient and attractive European Research Area.

Trans-national access (TNA) or virtual access activities, to support scientific communities in their access to the identified key research infrastructures.

Joint research activities (JRA), to improve, in quality and/or quantity, the integrated services provided at the European level by the infrastructures.

This book is one of the outcomes of the networking activities, where we focused on the best practices of protocols, standards, and ethics in pig research. In the call text of the European Commission, the words "harmonization", "standardization", and "use and deployment of global standards" were mentioned, referring to the ambition to create a more efficient research environment. The chapters in the book are co-authored by researchers from different laboratories and countries, indicative of the cooperation between research groups. The SOPs in this book complement the "Protocols for Standard Management and Recording in Pig Research Facilities" by Westin and Wallenbeck (2024), which were also developed during the PIGWEB project.

12.2 Materials and Methods and Standard Operating Procedures

Standard operating procedures describe in detail how certain research procedures should be carried out. The information of an SOP is summarized in the Materials and Methods section of a scientific manuscript, which, in theory, should ensure the repeatability of the research. If you randomly select a scientific article, would you be able to repeat the experiment under the same conditions as described in the article? I would not be surprised if the answer was "No" for most of the papers that you see.

The following text comes from the website of The Great British Bake Off (a TV series televised in many countries in the 2020s):

"This challenge separates the wheat from the chaff. Take one basic recipe, with the same ingredients and instructions, and ask our bakers to produce the finished product... sound easy? Well, any variation on the finished product will be a result of their own technical knowledge and experience—or lack of. Bakers are laid bare in this task, and this is where the pressure's really on in the Bake Off".

The same holds for the Materials and Methods section of many scientific articles: it provides "the basic recipe, with the ingredients and instructions", but, as in the TV series, it is likely that the outcome of the experiment depends on the technical knowledge and experience of the researcher. The caveat is in the word "basic", with limited specifics. We all know what temperature is (in baking or in an animal experiment), but where, when, and with what equipment was it measured? This does not mean that the outcomes of a published experiment are wrong or doubtful, but they are to be contextualized, and this context may not be sufficiently clear from the manuscript.

We now live in an era where sharing data becomes more important. Adopting the FAIR (Findable, Accessible, Interoperable, Reusable) data principles and the writing of a Data Management Plan has become standard practice. This illustrates the importance of going beyond the reporting of (average) results in a table of a scientific paper by providing access to the raw data. The same holds for the (summarized) materials and methods section in a paper: detailed operating procedures should be referenced and publicly accessible so that interested readers can consult them and be able to repeat the experiment under the same conditions.

In their paper, Hollmann et al. (2020) concluded that "the lack of reproducibility within laboratory research discourages successful implementation of the widespread adoption of research results in the scientific community. One way to improve it is to provide consistency and traceability of existing standards and laboratory practices that are achievable with precise and clearly written SOPs". The authors mention the "adoption of research results", which is very relevant for experimental animal research. Experimental animal research is under scrutiny by society, and it will be more difficult to carry out experimental animal research in the future than it is now. We will have to rely more on existing data from the literature, by re-analysing data from different studies through meta-analyses. In addition, to ensure that the data that we generate now has value in the future, it is our responsibility to provide a proper and traceable description of the methods and procedures that we use in animal research.

12.3 The Past and Future of Experimental Animal Research

Perceptions of experimental (animal) research change over time, and two examples will be given to illustrate this. In the first issue of the Journal of Animal Science, Benedict (1910) provided research suggestions for future animal nutrition research by highlighting the similarities between animal and human nutrition. He stated, "An ideal method of experimenting upon man would be to feed prisoners in the state institution a definite diet with a low protein content for a considerable period of time, possibly adding as a slight stimulant to adhere rigidly to the diet, the promise of a shortened sentence for hearty co-operation in the investigation. This while somewhat radical is not indeed beyond possibility, and is certainly worthy of consideration". In the 1960s, Lister and McCance (1967) carried out a series of experiments to see if growth was determined by time or by the state of the animal. They compared normally fed piglets after weaning with piglets that were feed-restricted for 1 year so that they would maintain their weaning body weight at 5.5 kg. After this 1-year period, they were offered feed ad libitum to monitor their performance.

I sometimes cite the Lister and McCance papers, not only to illustrate the change in ethics, but also to illustrate that growth is determined by time and state. But is it ethical to cite a paper that we now consider as ethically unacceptable? It is difficult to foresee future perceptions, but it is good to give it some thought and provide future generations of researchers with all the information they need so that they can judge if the research that we carry out now has value and is ethically acceptable to be reused.

We now consider experiments such as those of Benedict (1910) and Lister and McCance (1967) as unacceptable, but how will future generations look upon our current activities, 50 years from now? Many of us have been trained in a way that if we have a question, the answer is to be found in an experiment. Experimental animal research is a standard element in most animal science research projects (e.g. PhD project or collaborative national or international projects), and it is not the last resort if the answer cannot be found elsewhere. We publish around 14,000 peer-reviewed articles in animal and veterinary science journals every year, and it is not impossible that (part of) the answer we are looking for can be found there. Changing the way that we do experimental animal research probably requires a change in mindset, in how we train early-career scientists, and in the organization and use of research infrastructures, but these aspects are largely in our own hands.

References

Benedict FG (1910) Suggestions regarding research work in animal nutrition. J Anim Sci 1910:20–25. https://doi.org/10.2527/jas1910.1910120x

Hollmann S, Frohme M, Endrullat C et al (2020) Cost action CA15110. Ten simple rules on how to write a standard operating procedure. PLoS Comput Biol 16(9):e1008095. https://doi.org/10.1371/journal.pcbi.1008095

Lister D, McCance RA (1967) Severe undernutrition in growing and adult animals. 17. The ultimate results of rehabilitation: pigs. Br J Nutr 21:787–799. https://doi.org/10.1079/BJN19670081

Westin R, Wallenbeck A (2024) Protocols for standard management and recording in pig research facilities (march 6, 2024). Available at SSRN: https://ssrn.com/abstract=4750460

Index: List of ATOL-/EOL-identifiers from the SOPs

GPSR Compliance
The European Union's (EU) General Product Safety Regulation (GPSR) is a set
of rules that requires consumer products to be safe and our obligations to
ensure this.

If you have any concerns about our products, you can contact us on

ProductSafety@springernature.com

In case Publisher is established outside the EU, the EU authorized
representative is:

Springer Nature Customer Service Center GmbH
Europaplatz 3
69115 Heidelberg, Germany